Lecture Notes in Computer Science

Lecture Notes in Artificial Intelligence   14400

Founding Editor

Jörg Siekmann

Series Editors

Randy Goebel, *University of Alberta, Edmonton, Canada*
Wolfgang Wahlster, *DFKI, Berlin, Germany*
Zhi-Hua Zhou, *Nanjing University, Nanjing, China*

The series Lecture Notes in Artificial Intelligence (LNAI) was established in 1988 as a topical subseries of LNCS devoted to artificial intelligence.

The series publishes state-of-the-art research results at a high level. As with the LNCS mother series, the mission of the series is to serve the international R & D community by providing an invaluable service, mainly focused on the publication of conference and workshop proceedings and postproceedings.

Judit Bayer · Christian Grimme
Editors

Code and Conscience

Exploring Technology, Human Rights,
and Ethics in Multidisciplinary AI Education

 Springer

Editors
Judit Bayer 🆔
Institute for Information, Telecommunication
and Media
University of Münster
Münster, Germany

Budapest Business University
Budapest, Hungary

Christian Grimme 🆔
Information Systems
University of Münster
Münster, Germany

ISSN 0302-9743   ISSN 1611-3349 (electronic)
Lecture Notes in Artificial Intelligence
ISBN 978-3-031-52081-5   ISBN 978-3-031-52082-2 (eBook)
https://doi.org/10.1007/978-3-031-52082-2

LNCS Sublibrary: SL7 – Artificial Intelligence

This Springer imprint is published by the registered company Springer Nature Switzerland AG
The registered company address is: Gewerbestrasse 11, 6330 Cham, Switzerland

If disposing of this product, please recycle the paper.

This book is dedicated to all the students who pursue interdisciplinary studies between IT and social sciences.

Preface

1 Introduction

1.1 The Background of This Book - An Interdisciplinary Educational Project

This book represents the results of a groundbreaking project: an interdisciplinary, international, and interactive course on "AI, human rights and ethics: platform regulation and data protection", held in two consecutive semesters during the academic year 2021/2022. These were the second and further waves of the COVID-19 pandemic. Online teaching was first regarded as a constraint, but it was soon recognized to hold several new opportunities. Overcoming geographical barriers with the help of the online space made international cooperation easy also in the education sector. In a specific call by the DAAD, the International Virtual Academic Collaboration (IVAC) invited proposals that leverage these possibilities in education, by organizing blended, or purely online learning. The topic of the course was timely, and both students and lecturers demonstrated great interest. The content of this book has been created by students of this interdisciplinary course and edited by the co-teachers.

The conference papers included in this volume take the Springer Nature editorial policies into account.

1.1.1 The Purpose of the Interdisciplinary Course

The configuration of the course was responsive to the escalating demand within contemporary education for interdisciplinary approaches. This imperative is most pronounced in the domain of artificial intelligence (AI) education. The rapid and pervasive integration of AI applications across diverse sectors of existence, professions, and academic spheres underscores the essentiality of comprehending their mechanisms and associated vulnerabilities for all individuals pursuing scholarly enrichment. More than that, general rules and policy urgently need to be developed, and learned to be applied, so as to provide safety and trustworthiness and to protect human rights during the use of AI. AI applications have been initially developed exclusively by information technology (IT) professionals, without oversight by legal or ethical experts. IT professionals are frequently enthusiastic about the possibilities of technology, but at the same time, may downplay the social and individual risks presented by AI. Ethical and legal professionals have been increasingly involved in fundamental regulation of AI, but insufficient dispersion of technological knowledge has limited the capacity to develop appropriate solutions in all fields.

This course combined the disciplines of IT and law as its focus and professional content, and embraced even more faculties in its student cohort. The participating faculties were the Faculty of Information Systems at the University of Münster, Legal Studies at the University of Münster, Information Systems at the University of Twente, and Leiden

University. The participating students represented the subjects of Law, Computer Science, Information Systems, Media Technologies, Information Technologies, Business Administration, Data Science, and Data Management. The course was integrated into the English-language Master's programme in Business Information Systems (Information Systems) at WWU Münster, into the course offerings of the Institute for Information, Telecommunications and Media Law (ITM) at WWU Münster, and the curriculum of the University of Twente. In addition, the course was offered as a PhD course within the framework of the graduate programme of WWU Münster (6 ECTS). International credit transfer allowed us to open the course to any EU higher education institution in the second semester. The University of Leiden ensured flexible credit transfer to the Master's programmes in Business Informatics and Computer Science, as well as a PhD course. A continuation of the course is planned and longer-term funding is currently being sought. The leading co-teachers of the courses were PD. Dr.-Ing. Christian Grimme from the Faculty of Information Systems at the University of Münster, and Dr. habil. Judit Bayer from the Faculty of Law at the University of Münster. In both semesters, several lectures were held by external, partly international lecturers. The topic of the invited lecturers covered a spectrum from economics through law to computer science. The guest lecturers were all outstanding representatives of their disciplines, from foreign universities. In this respect, the course was not only able to contribute to the networking of the students with the lecturers, but also to surround the lecturers with a multidisciplinary approach to their topics.

1.2 Structure of the Course

Each of the two courses took one semester. After the first semester, evaluation and assessment were carried out, and slight improvements were applied for the second course. Both courses consisted of two parts: in the introductory part, six consecutive lectures were given by the co-teachers or the external lecturers. Sufficient time was devoted to discussion and answering questions.

The purpose of the introductory lectures was to introduce students to the technical background of AI, to the basics of human rights and ethical principles, and then to particular areas of AI governance and regulation, such as data protection; the basic structure of the draft AI Act; predictive policing; economic implications and democratic implications of AI.

The classes of the introductory part of the first course were structured as follows:

1. Course introduction (C. Grimme, J. Bayer, WWU)
2. Legal challenges, human rights challenges, ethical challenges. Data protection (J. Bayer, WWU)
3. Economic dimension (M. Ehrenhard, Univ. Twente)
4. Human rights and the Proposal of the European Commission for an Artificial Intelligence Act (M. Ebers, University of Tartu)
5. AI surveillance (L. McGregor, Essex Law School and Human Rights Centre)
6. How to use AI ethically in a good way (H. Hoos, Leiden University and COSEAL)

In the second semester, the teaching content of the course was somewhat expanded: first, the introductory session was expanded into two separate units to provide a bit more technical depth in each area (technology and law); second, two new topics were added: predictive policing and European organizational governance of AI. The final course content was as follows:

1. Introduction to the Course
2. Introduction to Artificial Intelligence (C. Grimme, WWU)
3. The Basics of Human Rights & Ethics (J. Bayer, WWU)
4. Enterprising with AI (M. Ehrenhadt, Univ. Twente)
5. Human Rights and the AIA-Proposal (M. Ebers, Univ. of Tartu)
6. Predict and Surveil - Data, Discretion and the Future of Policing (S. Brayne, Univ. of Texas)
7. AI Technology in Europe (R. Chavarriaga Lozano, Zurich Univ. of Applied Science, Head of Swiss office of CLAIRE)

After this "information transfer" part, the project part of the course started. Students were divided into groups with three to four members, where all students represented different disciplines. The student groups were required to hold at least two consultations, which had to be documented. The documentation allowed the teachers to follow the intellectual path taken by the students, to understand their decisions on the content, structure, cases, and literature used in the development of the manuscript. It also provided insight into the difficulties that students faced and how they overcame them. Together with the meeting documentation, the students submitted an outline which contained the main title, the subtitles of the planned paper, the literature, and some basic description of the planned content. Following submission of these, each group had at least one teacher-consultation, or more, if necessary, at which they received feedback on their outline. Then the students submitted a draft final paper, and a video presentation. At this stage, another amendment to the course was made in the second semester. The students got involved in **peer-reviewing** each other's papers. The peer-reviews were single blind: as it was impossible to anonymize the papers because students were aware of each others' projects, only the reviewers remained anonymous. Each student reviewed one paper: in this way, each paper received several student reviews and two teacher reviews. The reviewing task was supported with a guide provided by the teachers which explained the task and set out the ethical guidelines of reviewing. The review was a compulsory task, but not subject to assessment. Nonetheless, the students acted with enthusiasm, and made their reviews thoroughly. The reviews were forwarded to the authors, with the teacher's comments when necessary. Receiving concurring or dissenting reviews were equally educational for all students. Reading others' papers was also beneficial, especially for those students who were less experienced in research and academic writing. As the topics were different, plagiarism could not come into question.

Peer-reviewing proved to be an asset in the course and an element that the conductors would warmly recommend to other research courses as well. After the reviews, the students had the chance to improve their papers, and also their videos, if they wanted to. After the final submission, they were graded, and received further reviews and feedback for the purposes of publication. A majority of the papers (ten out of fifteen) were selected for this publication. The papers featured in this book are the first academic outputs of several students, and they often contain fresh ideas and perspectives. In one case, the course paper got published before this volume[1]. The grading also included assessment of the videos, coursework, and the progress made, based on the protocols.

1.3 The Relation Between the Two Semesters

After both semesters, a comprehensive quantitative and qualitative evaluation of the implementation could be carried out. The feedback after the first semester allowed us to further improve and develop the teaching units and didactic practices. Overall, the course was rated as "very good" by the students in both semesters. The major changes made in the second semester were as follows:

1. The teaching content of the introductory sessions was expanded and divided into two separate units to provide a little more technical depth in each area;
2. Peer-review was introduced to increase the cooperation dynamics of the research part of the course;
3. The international participation in the course was opened up. This brought in participants from several other EU Member States. However, the student dropout was more significant in these instances. This was at least partly due to the different semester schedules in the various Member States;
4. New didactic methods were employed to optimize virtual meetings, especially for additional interaction and discussion.

1.4 Evaluation of the Achieved Objectives

Important elements of the course were stimulating student interaction, continuous feedback, and an open learning and discussion culture that were maintained and expanded throughout the course. A variety of digital techniques (video conferencing via Zoom, collaboration platform via Google Classroom) were successfully used for connection. Despite the physical distance, it was possible to establish an orderly and engaged discussion culture in the course. The students interacted and discussed intensively with each other in the virtual environment. Breakout sessions were used to foster small-group discussion and interaction spaces, thus allowing thematic detailed discussions, with teacher support, when necessary. Then the discussions were summarized in the plenary session. A joint compilation of the developed ideas and insights thus became more efficient and structured. The virtual environment opened up new possibilities for content production for students. Thus, in the context of the second implementation, the use of self-produced student videos to present their results was applied as an innovative stylistic device which is to be used in further courses. The heterogeneous composition of the student groups for group work (different study areas and levels) enabled an interdisciplinary exchange in virtual collaboration.

---

[1] F. Koefer, I. Lemken, & J. Pauls (2022). Realising Fair Outcomes from Algorithm-Enabled Decision Systems: An Exploratory Case Study. In Proceedings of the International Workshop on Enterprise Applications, Markets and Services in the Finance Industry, pp. 52–67, Springer International Publishing, Cham.

1.5 Challenges

Educational challenges were found to be negligible. We attributed this to the novelty of the endeavour and the enthusiasm of students (and of teachers). The challenges experienced were mainly of an organisational nature. In particular, the lack of synchronization of semester times (e.g. summer semester in Germany, spring term in the Netherlands) posed a challenge in running the courses for both students and administrators. The associated administrative workload within the curricula and examination offices highlights a problem of international cross-curricular teaching that has not yet been conclusively solved. This synchronisation could be enhanced by strategic planning by participating universities, alternatively by allowing more flexibility for international cross-educational courses. A long-term insertion of the programme into the curriculum of the participating universities was seen as desirable, however, the inflexibility of the curricular design did not support the realisation of this plan.

During the second implementation, quite a few of the registered participants dropped out of the course. Some participants remained passive participants who attended the course out of interest and were not interested in acquiring a certificate of achievement. This gave us a signal that opening the call to an unlimited audience may bring about unnecessary complications, and overburdening of the public educational system. In the second term, due to the lower number of participants and the unequal representation of the students' disciplines, as well as due to mid-course dropouts, some group memberships decreased to only two. This made group work less efficient, and also limited the effects of interdisciplinarity.

1.6 Conclusion and Perspectives

The first and second implementation of the course within the framework of the funding measure was deemed successful from both the lecturers' and the students' point of view. In particular, the great student interest in the course shows that such interdisciplinary programmes are in demand. Of more than 92 students who expressed interest in the course, 45 participants were admitted in the first round and 40 students in the second round (in each case about twice as many participants as planned in the application).

The results of the two experimental semesters strongly confirm the idea that offering interactive education at the crossroads of artificial intelligence (AI) and legal studies is crucial. It would clearly present a valuable educational opportunity that yields reciprocal advantages for students coming from various academic domains. Our changing landscape emphasizes the need to educate a generation that understands the ethical dimensions of AI technology.

Additionally, such international educational collaborations in the EU are key for creating educational synergies. This would enhance competitiveness at both industrial and academic level.

November 2022   Judit Bayer
                                                               Christian Grimme

Acknowledgements

The editors of this volume, Judit Bayer and Christian Grimme, thank the German Academic Exchange Service for its funding of the IVAC project "AI, Human Rights, and Ethics". Christian Grimme was also supported by the European Research Center for Information Systems (ERCIS) and part of the joint research project HybriD, which was funded by the German Federal Ministry of Research and Education. The course would not have been possible without the contribution of the invited lecturers, Michael Ehrenhadt, Martin Ebers, Sarah Brayne, Ricardo Chavarriaga Lozano, Lorna McGregor, and Holger Hoos.

Acknowledgements

Contents

List of Contributors

Richard Albrecht Department of Information Systems, University of Münster, Münster, Germany

Vasos Arnaoutis University of Twente, Enschede, The Netherlands

Ilse Arwert Leiden University, Leiden, The Netherlands

Giulio Barbero Leiden Institute of Advanced Computer Science, Leiden University, Leiden, The Netherlands

Charlotte Daske Department of Information Systems, University of Münster, Münster, Germany

Adelson de Araujo Faculty of Behavioural, Management and Social Sciences, University of Twente, Enschede, The Netherlands

Kian Deutz Department of Information Systems, Westfälische Wilhelms-Universität Münster, Münster, Germany

Michaël Grauwde Leiden Institute of Advanced Computer Science, Leiden University, Leiden, The Netherlands

Marie Griesbach Department of Information Systems, Westfälische Wilhelms-Universität Münster, Münster, Germany

R. N. Guérin Leiden University, RA, Leiden, The Netherlands

Charlene Hinton University of Münster, Münster, Germany

E. I. S. Hofmeijer Faculty of Electrical Engineering, Mathematics and Computer Science, University of Twente, AE, Enschede, The Netherlands

L. M. Kester University of Twente, AE, Enschede, The Netherlands

Leve Lorenzen University of Münster, Münster, Germany

Florian Medert Department of Information Systems, University of Münster, Münster, Germany

Amelie Mehlan University of Münster, Münster, Germany

Dominik Neumann Department of Information Systems, Westfälische Wilhelms-Universität Münster, Münster, Germany

Jan Fridtjof Otto Faculty of Geo-Information Science and Earth Observation, University of Twente, Enschede, The Netherlands

Mária Magdaléna Perczel Faculty of Law, Eötvös Loránd University of Sciences, Budapest, Hungary

Janina Pohl Department of Information Systems, Westfälische Wilhelms-Universität Münster, Münster, Germany

Pradipa P. Rasidi Department of Information Systems, Westfälische Wilhelms-Universität Münster, Münster, Germany

Jeroen G. Rook University of Twente, Enschede, The Netherlands

Fareeha Saeed Westfälische Wilhelms-Universität Münster, Münster, Germany

Alireza Samiei The Leiden Institute of Advanced Computer Science, Leiden University, Leiden, The Netherlands

Lennart Schmidt Westfälische Wilhelms-Universität Münster, Münster, Germany

L. W. Sensmeier University of Münster, Münster, Germany

Ajith Sivakumar Department of Information Systems, Westfälische Wilhelms-Universität Münster, Münster, Germany

Syeda Amna Sohail Department of Data Management and Biometrics (DMB), Faculty of Electrical Engineering, Mathematics and Computer Science (EEMCS), University of Twente, Enschede, The Netherlands

Marise van Noordenne Leiden Institute of Advanced Computer Science, Leiden University, Leiden, The Netherlands

Lieke van Zijl Leiden University, Leiden, The Netherlands

Julian von Lilienfeld-Toal Department of Information Systems, Westfälische Wilhelms-Universität Münster, Münster, Germany

Janek Wenning University of Münster, Münster, Germany

Kennet Winter Department of Information Systems, Westfälische Wilhelms-Universität Münster, Münster, Germany

Acronyms

ABC	American Broadcasting Company
AI	Artificial Intelligence
API	Application Programming Interface
ATM	Automated Teller Machine/Automatic Teller Machine
CCTV	Closed-Circuit Television
COVID-19	Coronavirus Disease 2019, DAAD - Deutscher Akademischer Austauschdienst
DPA	Dutch Data Protection Authority
DPA	Deutsche Presse Agentur (German press service)
EDPB	European Data Protection Board
EEG	Electroencephalogram
EEOC	Equal Employment Opportunity Commission
EIJI	Edinburgh International Justice Initiative
EU	European Union
FaceID	Face Identification
FACT	fair, accurate, confidential, and transparent (principles)
FAIR	findable, accessible, interoperable, and reusable (principles)
FRT	Facial Recognition Technique
G20	Group of 20 (economically largest countries)
GDPR	General Data Protection Regulation
IEEE	Institute of Electrical and Electronics Engineers
ISO	International Organization for Standardization
KPI	Key Performance Indicator
LIDAR	Light detection and ranging
LSTM	Long short-term memory (usually: neural network)
ML	Machine Learning
NASDAQ	National Association of Securities Dealers Automated Quotations
NIAH	Hungarian National Authority for Data Protection and Freedom of Information
OECD	Organisation for Economic Co-operation and Development
OTC	Over-the-counter (trading)
PIN	Personal identification number
SAR	Socially Assistive Robot
SEC	U.S. Securities and Exchange Commission
SME	Small and medium enterprises
UI	User Interface
US	United States (of America)
VR	Virtual Reality

Facial Recognition in the Public Space: Challenges and Perspectives

Ilse Arwert[1], Amelie Mehlan[2(✉)], Jeroen G. Rook[3], and Janek Wenning[2]

[1] Leiden University, Leiden, The Netherlands
[2] University of Münster, Münster, Germany
{amelie.mehlan,j_wenn06}@uni-muenster.de
[3] University of Twente, Enschede, The Netherlands
j.g.rook@utwente.nl

Abstract. Due to technological advances, AI systems, among them facial recognition, are becoming more commonplace, a development which garners ethical and privacy-related concerns and has prompted the European Commission to develop the proposed AI Act of 2021. This study investigates whether current legislation, including the proposed AI Act, is enough to properly regulate the use of facial recognition systems in public spaces in the European Union. To this purpose, we outline the current status of EU legislation regarding these systems, examine several case studies of real-world use of such systems and the response thereto, and identify overarching ethical and legal issues that arise from these case studies. We find that currently, the ambiguous phrasing in the proposed legislation, combined with lack of enforcement of the need for consent, means that legislation is not sufficient to regulate the use of facial recognition in public spaces in the European Union. Therefore, based on the observations from the case studies, we make several recommendations that, when followed, encourage the use of facial recognition systems in public spaces in an ethical, legal and responsible way.

1 Introduction

Biometric systems are being developed and improved with unmistakable speed. Technological functionalities and possible use cases of available systems are maturing and becoming more diverse, as such biometric systems become more accessible to the general public. With biometric systems, we refer to systems that recognize, identify or verify the identity of individuals based on physical characteristics, usually through machine learning. Facial recognition is an exceptionally risky biometric system because facial images can easily be captured without consent, with commonly available security cameras. Faces are easier to capture in public and without the subject's awareness or consent than other common biometrics such as teeth profiles or fingertips. Faces are often uncovered, and even relatively low-resolution footage can be clear enough to be used for facial recognition.

J. Bayer and C. Grimme (Eds.): AI, Human Rights and Ethics, LNAI 14400, pp. 1–16, 2025.
https://doi.org/10.1007/978-3-031-52082-2_1

The use of facial recognition (and other biometrics) in automated systems has been the subject of debate for many years. It is seen as a stepping stone to mass surveillance systems and may give rise to many privacy-related and other ethical concerns [1]. Within the European Union, these concerns have led to the development of legislation regarding the use of biometrics. Examples of such legislation are the GDPR and the European Commission's proposal for the Artificial Intelligence Act [2]. However, legislative powers may not be fully informed regarding the most recent developments within the field, and therefore face difficulties in assessing the efficiency and implications of such systems and their future potential.

Much has been written about biometric systems as a legal and political problem, such as Sprokkereef and Hert's contribution to the 2012 book Biometrics, Privacy and Agency, in which they examined the legal challenges associated with biometric systems. The main risks of facial recognition, according to them, are invisible data collection, data protection regulation and the dangers of profiling - all issues that are still relevant today [3]. However, the speed at which the field of facial recognition identification is developing means that any writings are soon outdated. Additionally, as this technology is becoming increasingly accessible to private individuals or other nongovernmental entities, new concerns arise due to a lack of oversight.

The 2021 release of the proposed AI Act by the European Commission inspired us to take a closer look at the current status of legislation regarding facial recognition systems. We were unconvinced that the proposed AI Act provides accurate regulations to cover present-day use of facial recognition, let alone future developments and uses. This leads us to our research question: is current legislation enough to properly regulate the use of facial recognition systems in public spaces in the European Union?

In this paper, we outline the current legal framework for dealing with facial recognition technology in the EU, examine several case studies of past and current uses of facial recognition systems and finally provide recommendations to ensure that the use of such systems is ethical, legal, and proportional.

2 What is Facial Recognition?

Humans have many physical characteristics that distinguish us from one another, called 'biometrics'. Some of these biometrics, like fingerprints, teeth profiles and faces, are so unique to individuals that they can be used to identify persons. Consequently, biometrics have been used in forensic investigation and security systems [4]. Biometrics are especially interesting for security systems, as physical features are harder to steal or replicate than a key card or password. This makes biometrics a tempting security measure for public or private entities that require restricted access to facilities or information.

One difference between faces and the other aforementioned biometric traits is that faces are also used by humans to identify each other. Additionally, faces are harder to conceal in public without calling attention to oneself. Another important difference is that if you take a picture of someone, it is likely that 1) their face is included, and 2) the resolution is good enough that their face can be used for identification. Fingerprints, teeth profiles et cetera cannot be extracted from publicly available footage as easily.

In broad terms, we define facial recognition as the automated detection of a face and the further analysis of a face.

The sources used for facial recognition are usually images or video captures, but other sources such as three-dimensional scans (e.g. Apple FaceID) or thermal imagery can also be used, and the light spectrum that is captured can vary (e.g. infra-red or visible spectrum). All these different sources are collectively referred to as captures in this paper.

We define four different degrees of facial recognition: detection, attribute recognition, face matching, and identification.

Face detection is concerned with determining whether there are faces present in a capture and, if there are, isolating them.

Attribute recognition or soft biometrics involves classifying characteristics (i.e. gender or age) based on a face [5]. Some studies use facial recognition to determine the emotional state of a person [6]. It is questionable to what extent there is a causal link for these classifications, and creates a danger- ous precedent to use faces for other predictions, such as the likelihood of a person committing a crime in the future (an example with a high danger of perpetuating systemic racial bias) [7].

Face matching assesses the similarity of two face captures. An example: to com- pare a face with the faces that are allowed to unlock a smartphone, or to com- pare a face capture to a database of faces. Another example of face matching is in large-scale CCTV applications, to match captures from one camera to footage of other cameras in order to capture movement patterns of individ- uals. The person the face belongs to can remain anonymous, by removing identifiable face attributes in the stored face captures [8,9].

Personal identification is the most specific, intriguing and potentially dangerous use of this technology. Here, the input of the face is used to uniquely identify a person by connecting someone's face to other data sources such as the person's name and address, criminal records or personal interests.

2.1 Facial Recognition Methods

Machine learning describes the process of feeding data into a series of algorithms that can automatically learn representations [10]. Deep learning methods are a popular subclass of machine learning and are nowadays particularly used in facial recognition systems to verify or identify a person's identity through analysis of their facial features.

Verification involves comparing a person's face with a previously collected gallery of their face, such as the face verification system which is used on many new phones to unlock the device. This falls under the third degree of facial recognition: face matching. In the case of identification, the face is compared to a large database of labelled persons [11]. This falls under the third or fourth degree, depending on how much information such a database contains: if it only has the faces itself, it is face matching; if there is personally identifiable information, it is personal identification. These deep-learning-based systems are capable of recognizing faces and extracting facial features.

To create a high-precision deep learning facial recognition model, a lot of training data is needed, often with several captures per person. One approach to train a model is to teach it to filter out the facial features through many calculations in an artificial neural network, and finally to output a feature vector. This feature vector is then compared to the feature vector of another facial image of the same person and to that of a different person. The neural network's parameters are then adjusted to minimize the distance between the vectors of the same person without decreasing the distance to the vector of the other person. Ideally, different images of the same person will be close together (seen as similar) without being close to images of any other person. A well-trained model will then be able to precisely match a new facial image to the corresponding person saved in a database.

These systems naturally require data processing: from the data required to train the models, to the saved feature vectors for comparison, to the capture and analysis of the facial images themselves. This, combined with privacy concerns, means that legislation is needed that outlines the rules of using and creating such systems. Below, we outline the current state of this necessary legislation.

3   EU Legislation

Currently, there is no legal framework in the EU that deals specifically with facial recognition software or biometrics in general. Instead, legal regulations for this area can be found in three sets of rules: 1) the general fundamental rights, which represent elementary rights of defence for citizens against the government, as well as between citizens; 2) EU data protection law; and 3) the Artificial Intelligence Act proposed in 2021 by the European Commission (hereafter 'AI Act') [2]. In the following section, we provide an overview of the current European regulatory framework.

3.1   General Fundamental Rights

The Charter of Fundamental Rights of the European Union enshrines various fundamental rights in primary EU law enjoyed by EU citizens and residents. Almost all of these fundamental rights can conceivably be endangered by the use of AI systems [12]. One example is the right to freedom of assembly and of association provided by Art. 12 EU Charter.

Art. 12 (1) EU Charter states that 'everyone has the right to freedom of peaceful assembly and to freedom of association at all levels, in particular in political, trade union and civic matters, which implies the right of everyone to form and to join trade unions for the protection of his or her interests'. These rights can be directly restricted by the use of facial recognition software in public spaces when used for tracking purposes.

As an example, some Australian states have built-in real-time facial recognition software in their security cameras [13]. This technology might have been used when the police tried to identify protesters that protested the ongoing COVID-19 lockdown [14]. Such surveillance can not only directly impact general fundamental rights but also cause a chilling effect. In a legal context, a chilling effect[1] is the discouragement or inhibition of the legitimate exercise of natural and legal rights by the threat of legal sanction. The effect has notably been emphasized by the Committee of Ministers of the Council of Europe, observing that surveillance and digital tracking technologies can have a detrimental chilling effect on citizen participation in political, cultural and social life [15].

In addition, it is conceivable that further rights are affected, e.g. the right to fair trial (Art. 47 EU Charter) or the dignity of the human person (Art. 1 EU Charter) [12]. The usage of facial recognition by law enforcement authorities is characterized by a significant degree of power imbalance and may lead to surveillance, arrest, or deprivation of a natural person's liberty (Recital 38 AI Act). This can have an impact on the right to a fair trial, which guarantees everyone equal opportunities, equality of arms and effective means of defence.

3.2   EU Data Protection Law

The European data protection law is relevant regarding the data that's being used with facial recognition systems. The General Data Protection Regulation (GDPR) explicitly regulates the use of biometric data. Art. 9 (1) GDPR considers the processing of biometric data for the purpose of uniquely identifying a natural person as the processing of a special category of personal data, or 'sensitive data', which shall in principle be prohibited. But not all usage of biometric data falls under the GDPR. The European Data Protection Board (EDPB) has emphasized that therefore, the definition in Article 4(14) GDPR has to be taken into account, for which three types of criteria must be considered: the nature of data (data must relate to physical, physiological or behavioural characteristics of a natural person), the means and way of processing (data must result from a specific technical processing) and the purpose of processing (data must be used for the purpose of uniquely identifying a natural person) [12,16].

For this reason, the EDPB has recognized that video footage of an individual cannot in itself be considered processing of biometric data in the sense of Article 9 GDPR (idem), and that 'when the purpose of the processing is for example to distinguish one category of people from another but not to uniquely identify anyone the processing does not fall under Article 9' [16].

As recently stated by González Fuster and Nadolna Peeters, that means that it is a case of processing biometric data within the meaning of EU data protection law 'if the data allows or confirms the unique identification of a natural person,

---

[1] Definition taken from: https://www.yourdictionary.com/chilling-effect.

who is a person identified or identifiable, and the processing takes place for the purpose of uniquely identifying that natural person' [12].

In cases where facial recognition is used for automated decision-making, in addition to Art. 9 GDPR, Art. 22 GDPR can be affected. This states that individuals should have the right not to be subject to a decision based solely on automated processing, including profiling, which produces legal effects concerning them or similarly significantly affects them. This seems like a great protection against the risks inherent in the use of facial recognition, but critics argue that the 'normative radius of Art. 22 GDPR is more limited than its heading (...) suggests' [17]. The reason for this criticism is that the wording of Art. 22 GDPR limits its effects to legal and similar consequences.

3.3   Artificial Intelligence Act

In April 2021, the European Commission proposed the so-called AI Act, in an attempt to regulate AI systems uniformly and to set clear limits for them. Art. 5 draft AI Act prohibits the use of so-called AI systems with an 'unacceptable risk'. Art 5 (1) (d) AI Act states that 'the use of 'real-time' remote biometric identification systems in publicly accessible spaces for the purpose of law enforcement is classified as one of those unacceptable risks'. Nevertheless, it makes exemptions for the search for specific potential victims of crime, for the prevention of specific threats such as a terrorist attack and for the detection, localization, identification or prosecution of perpetrators or suspects of a criminal offence.

According to Art. 5 II AI Act, further prerequisites are linked to the probability and extent of the occurrence of damage, and the consequences for the rights and freedoms of the persons concerned. It also mentions unspecified protective measures and conditions with regard to time, space, and persons concerned, including a prior decision by a court or an authority. However, this legislation still opens the gateway for further far-reaching surveillance.

It seems the Commission is making efforts to dispel doubts about state use of biometric identification systems. This is supported by the fact that biometric systems are mentioned in Art. 5 AI Act among prohibited practices. But with the ambiguity and breadth of current exemptions, there essentially is no ban on biometric identification systems at all. The requirements in the proposed AI Act are far from sufficient to allay concerns about the risks; not only to the individuals concerned, but to society as a whole. Critics have noted that the permitted purposes for using such a system are too vague; with the current wording, suspicion of criminal intent is already sufficient to allow biometric identification. The protective measures to be taken are hardly specified; as is well known, the court order often stays without effect. [18].

However, further discussions are led regarding the ban of facial recognition in the EU. At the time of closing this manuscript, the text of the AI Act is still developing, and future legislative propositions may be forthcoming as well.

4   Case Studies

In this section, we provide some case studies with real-world examples of the use of facial recognition systems in public spaces in the European Union. For each, we will include a description of the intended function of the system, public and legal reactions to the use of the system, and a brief ethical discussion of the system. This last part will be expanded on in Sect. 4.2, where we identify patterns that arise from these case studies.

4.1   Examples

4.1.1   Use of Facial Recognition by Swedish and Finnish Law Enforcement

There is an increasing number of facial recognition software uses by law enforcement in the EU. A highly controversial example was the case where the company Clearview AI[2] was involved. Clearview AI was founded by Hoan Ton-That and Richard Schwartz in the United States and offers a facial recognition system to law enforcement agencies. Users download an application on a device and upload an image which is biometrically matched against a large database with images from e.g. news media, websites and social media like Facebook, Instagram, or YouTube. The facial recognition system then sends the source links of matched faces to the user. After Clearview AI had a data leak, their customer list became public, bringing some European customers to light.

One of these customers was Swedish law enforcement. They were heavily criticized when it came to light that a number of individual police officers used the software without prior authorization for operational activities, including identifying complainants in suspected sexual offences against children, and in reconnaissance activities to try to identify unknown persons in serious organized crime. The Swedish Authority for Privacy Protection imposed an administrative fine of approximately €250,000 on the Swedish Police Authority, as they were unable to implement appropriate organizational actions to ensure and prove that the processing of personal data in this case was carried out in compliance with the Criminal Data Act. Furthermore, the Police Authority was required to make sure to the highest possible extent that all data which was sent to Clearview AI was fully erased [19].

A similar case happened in Finland, where a unit of the National Bureau of Investigation that specializes in the prevention of child sexual abuse had used a trial version of the Clearview AI facial recognition software to identify potential victims. The decision to try the software (in other words, the decision to process personal data via a third party) was made independently, without the knowledge of the National Police Board. Moreover, the police started this experiment without knowing how the personal data was processed on the side of Clearview AI, or how long the data would be stored. The Office of the Data Protection Ombudsman ordered the National Police Board to notify all involved subjects

---

[2] https://www.clearview.ai/.

of the incidents and to make sure that all data that was sent was erased by Clearview AI [20].

4.1.2 Dutch Facial Recognition by Non-Governmental Entities

In 2019, it was revealed that a Dutch supermarket had connected the cameras at its entrance to a facial recognition system in order to compare customers' faces with those of people who had been banned from entering stores after shoplifting incidents. Following media coverage, the store was contacted by the Dutch Data Protection Authority (DPA) with a request for information. Two days later, the store disabled the technology. However, the owner did express an interest in re-enabling the system, upon which the DPA issued a formal warning to the store in January 2020. At that point in time, facial recognition was banned in all but two situations. The first exception was when the people being filmed had provided explicit consent for their data to be processed. This was not the case for the store, as they did not provide clear information to their customers that their faces were being analysed. The second exception is in the interest of substantial public interest. The store argued that keeping out shoplifters should count for this, but the DPA disagreed, citing the only example provided in the law: ensuring the security of a nuclear power plant. As preventing a nuclear event is several magnitudes more important to public safety than preventing shoplifting, they declined the store's usage of facial recognition technology [21].

However, similar cases exist in the Netherlands that did not face such backlash. People celebrating Carnival in a popular partying street in 2019 were being filmed, and their faces scanned and compared to those of people who had been banned from the street after misbehaving on prior nights. After misbehaving, people would be removed from the street and were issued a warning: if they returned within three days, they would be recognized and removed by security. However, 'regular' party-goers were not informed that their faces were being scanned and compared with those of these misbehaving guests [22]. Attempts by news companies to contact Profi-Sec, the security company responsible, failed; as soon as reporters mentioned facial recognition, they were shut down [23]. No formal warnings have been issued against Profi-Sec or its use of facial recognition; evidently, misbehaving party-goers counted as enough of a threat that the use of facial recognition was allowed.

Similar cases have been seen in casinos (notably, one casino disabled its system after public backlash even though it was in accordance with legislation), Schiphol Airport and football stadiums, among others. The EDRi network's new independent report by the Edinburgh International Justice Initiative (EIJI), published in 2021, outlines a number of these examples, and concludes that these uses of facial recognition are incompatible with the EU Charter of Fundamental Rights, the European Convention on Human Rights and the GDPR. Furthermore, it states that these uses of facial recognition are out of proportion compared to the actions they wish to reduce, and therefore cannot be justified under current legislative guidance. Despite their extrajudicial nature, though, the use of these systems is largely going unpunished, as the DPA does not have the capacity to enforce privacy laws at the necessary scale [24].

4.1.3   Large-Scale Surveillance in Hungary

Another worrying example is that of the Hungarian mass surveillance project known as Szitakötő ("Dragonfly"). Note: there appears to be no relation with Google's controversial data analysis search engine project of the same name, which was discontinued amidst privacy concerns in 2018. The Hungarian mass surveillance project will be referred to as Project Dragonfly throughout this section.

In 2018, the Hungarian government announced a plan for a centralized CCTV system, with data stored in one location. This project would involve the installation of 35 thousand cameras in Budapest and across the country, which would be connected to a facial recognition system. In addition to this, parking meters would be equipped with wide-angle cameras monitoring pedestrians. The central data centre would also store recordings from road traffic cameras, ATMs, and banks. Several news sites mention concerns regarding data protection and privacy related to this system. For one, recorded data would be preserved for 30 d instead of the previous maximum of five days. Secondly, data would be collected *en masse*, and could be used to create a complete profile of any person by tracking their face across the database [25–27].

Little is known about the current state of Project Dragonfly, due to an overlap between private and public sectors regarding the maintenance and development of this project, as well as the complex relation between Dragonfly and other information systems. The Hungarian government has received numerous warnings regarding data protection and the right to privacy from both public and private organizations, such as the Hungarian National Authority for Data Protection and Freedom of Information (NAIH) [28]. However, there is a striking lack of social debate around the current state of human rights and privacy in Hungary when it comes to this project, and a similarly striking lack of transparency regarding the current status of Project Dragonfly. In December 2019, the Hungarian Parliament passed a number of amendments legalizing the use of forensic and live facial recognition technology by the Hungarian Police. If an individual apprehended by the police cannot present an ID document, the police agent can upload a photograph of said individual, which can be instantly verified against the national registry of citizens by a facial recognition algorithm - something which is only possible due to the central data collection system which is part of Project Dragonfly [29].

4.2   Overarching Conclusions Risk Assessment

In this section, we will discuss patterns we noticed in the discussed case studies, ranging from ethical concerns to systematic issues with regard to current legislation and the enforcement thereof. It is important to keep in mind that these issues are often not limited to facial recognition, but could be generalized to a wider context (multiple types of biometric systems, or even as wide as all AI systems).

In the case studies of facial recognition discussed above, the ethical aspect plays a central role, especially as an interplay between the privacy of an individual and the security of society, protected by the EU values related to privacy, for instance the value of freedom, which mentions "respect for private life", or the value of human rights, which mentions the "right to the protection of personal data" [30]. Using a facial recognition system without proper insight into how it processes its data is far from being respectful to a citizen's private life; recall the aforementioned Clearview AI example, which browses through the internet to find every image containing someone's face for recognition purposes, used by law enforcement without permission or insight into the back-end of the system. Nevertheless, we recognize that there are some situations in which the use of such systems is ethically justifiable, for instance when national security is in danger or human lives are at stake. But where is the line between these security contexts?

Legislation is ambiguous; take for example the exceptions to the ban on the processing of biometric data. What counts as an 'important enough' safety issue is often up for interpretation, as argued by the Dutch supermarket. Even with the AI Act, there are exceptions, such as for the detection, localization, identification, or prosecution of perpetrators or suspects of a criminal offence. Even a suspicion is enough to wave away the ban on processing biometric data, essentially giving law enforcement agencies free rein on the use of facial recognition systems. The ambiguous and/or lenient state of current legislation enables disproportional use of such systems, which are unneeded invasions of privacy even when they are technically legal. This combines with the inability of data protection agencies to keep up with the unlawful use of facial recognition systems. Too many public or private entities see the potential use of these systems and initiate their use; the data protection agencies simply cannot keep up. In this deluge, even some illegal systems are allowed to stay active, simply because they are not spotted by data protection agencies or because those agencies do not have the resources to enforce their termination.

Another contributing factor to the lack of enforcement is the lack of transparency regarding the use of facial recognition systems in public spaces - another common thread in the case studies above. Law enforcement agencies, security companies or private individuals who employ these systems do not have an incentive to advertise their use, and often use facial recognition systems without the authorization to use them. The public are therefore often not aware that these systems are being used, which also means they cannot consent to it - a clear violation of the GDPR. This could be partially remedied by notifying the public that a facial recognition system is being used in a certain location, for example by placing stickers informing people of this. However, when the public is informed that facial recognition systems are in use, especially by nongovernmental entities, this often results in public backlash. Negative media coverage and risk of a negative reputation have in the past resulted in companies terminating the use of these systems, even when they did not face legislative consequences for their use. This has even happened in cases where the use of facial recognition technology was perfectly legal [23]; public backlash is often based on not understanding the system and gut feelings rather than fact. At the same time, the lack of information about the reasons for using the system makes consent invalid, and thus makes deployment of the system unlawful.

In conclusion, several consistent issues are clear after examining these case studies. Firstly, facial recognition systems need to be clearly marked when they're being used to ensure that the public can consent. Secondly, to ensure that this consent (or lack thereof) is based on more than a gut feeling, the public needs to be informed about what personal information is processed, and how this might impact their privacy. Thirdly, to enforce this requirement for consent and to reduce the presence of illegal facial recognition systems, data protection agencies need to have the resources to enforce this. Fourthly, we must reduce the ambiguity in current legislation vis-à-vis versus exceptions to the ban on processing biometric data, to clearly demarcate which systems are legal - and which are not.

5   Looking Ahead Recommendations

Based on the case studies from Sect. 4.1 and their overarching conclusions presented in Sect. 4.2, we have five recommendations for the usage of and dealing with facial recognition in public spaces:

Decomposition Facial recognition systems are multifaceted; therefore, ethical concerns cannot be blamed on one part of the system alone. To critically assess where the most pressing ethical concerns lie, we decompose these systems into three parts; sources, technology, and context. Sources cover the data (that has biases [31]) where the systems are trained with, technology covers the working hardware and software, and context covers the setting and goal the technology is deployed for. For legislation specifically, we recommend that for each of these parts guidelines and recommendations are given independently of each other. For example, facial recognition is often closely associated with mass surveillance systems and their threats and deficiencies are susceptible to be blended with the technology. Ruling out a technology based on the ethical concerns for one of its use cases is not fair and holds the technology back from being used for good purposes. For example, arguably, facial recognition in a multi-tracking multi-camera setting [32] can thrive for good in the field of urban computing [33,34], by providing detailed bottom-up crowd flow tracking data. In an isolated setting, without a context, facial recognition technology is neither ethical nor unethical. However, one should always aim to minimize (ethical) expense of data and technology to achieve the goal of the context. I.e., if there is a technology that is less intrusive compared to facial recognition, that should be used.

Societal impact The possibility of being remotely identified and that personal information can be linked to that identification impacts societal behaviour [35]. This is described as the <u>surveillance cue effect</u> and can be considered as a chilling effect (Sect. 3.1). People exhibit more socially desired behaviour in the presence of others and having the perception of being watched does this as well [36], but does not further increase socially desired behaviour. Simultaneously, the necessity of constantly watching each other can create an artificial feeling of insecurity and also in the feeling that you are not trusted by the authorities [37]. Since trust is bidirectional, this also implies that citizens are less likely to trust these authorities [37]. This especially applies to mass surveillance systems. In our opinion, reduced trust in each other supersedes the positive social effects of facial recognition.

Bodily integrity People cannot easily hide or replace their face in order to refrain from being subjected to facial recognition. This is one of the reasons faces are a special category of personal data in the GDPR. We consider this an intrusion of bodily integrity [38] – even when no facial information is stored. Consequently, we recommend that, whenever possible, alternative, less intrusive technologies and/or systems should be used.

Transparency Facial recognition exists and will not go away. Even with a prohibition on the technology, people and organizations would continue researching and developing it - albeit behind curtains, in private labs. This would give more room to potentially malicious actors. Therefore, we assess that a hard ban on facial recognition is unfavourable, and its research and development should be encouraged in the scientific community. This will help to keep its development transparent, and to set ethical and legal requirements.

Enforcement We recommend moving from post-check to pre-check. In many cases, facial recognition was deployed in systems where afterwards it was concluded that these systems violated one or multiple laws. This does not only affect public opinion toward the organizations that made or used the system, but ripples out into public opinion on the use of this technology in general. This harms the trust people have in facial recognition, and, in general, in AI. To avoid this, we advocate that systems are reviewed on their legality, their technological robustness, and their societal impact, both in the initial design of the system and upon completion of the system. Such procedures and checks already exist in many other engineering fields. For example, when a bridge is built, permits need to be acquired, the structural robustness needs to be calculated and assessed once it is completed, and environmental impact studies need to be conducted. We argue that these assessments need to be conducted for the complete systems (sources, technology, and context). This contrasts other work [39] that argues to regulate data and algorithms separately. Additionally, since creating and using AI-systems requires a high degree of knowledge in order to obtain robust, trustworthy, and unbiased solutions, we also recommend having some kind of certification for the organizations that build and use them, comparable to ISO certification.

The presented guidelines are not limited to facial recognition only and do extend to AI systems in general. This especially holds for the guidelines <u>decomposition</u>, <u>enforcement</u>, and <u>transparency</u>. There are also recommendations that seem to conflict with each other, like decomposition and enforcement. For the decomposition guideline we argue to assess the ethics and concerns of a system independently of each other while in the guideline enforcement we argue that systems need to be evaluated as a whole. We do note that these two suggestions are not mutually exclusive: decomposition to identify in which component the ethical concerns lie and enforcement as a full-stack assessment.

In summary, these recommendations aim to increase the trust in facial recognition technology such that it can be used for good purposes and to obtain a better public acceptance of the technology. We also acknowledge that facial recognition technology is powerful and intrusive and therefore should only be used when it is a proportional solution to the problem it is meant to solve and performs better than alternative technologies.

6   Conclusion

We investigated facial recognition in European public spaces from multiple perspectives. We examined different degrees and methods of the technology, the legal frameworks that are involved with the technology, and we derived five guidelines from multiple use cases where facial recognition was used. By drawing conclusions from all these perspectives, we are now able to answer our research question: is current legislation enough to properly regulate the use of facial recognition systems in public spaces in the European Union?

We conclude that current legislation in the EU recognizes that the privacy of all citizens is a particularly valuable right. It will be a major task to reconcile this right with the pursuit of innovation. Under no circumstances should the legislature impose overly strict regulations in this regard. This would entail the risk that new technologies would no longer be developed in the EU at all, which would weaken the EU as a business location. In addition to this, it would make the security of these technologies even more questionable and difficult to oversee. As with other new technologies, e.g. genetic engineering or new medical products, the social added value of facial recognition must be recognized and weighed against the risks. As with these technologies, the legislature must be guided by the findings of science and make the necessary provisions to implement its directives. Currently, the biggest problem seems to lie in the control and enforcement of existing rules. As various current proceedings under the GDPR against Google or Meta show, the EU is not succeeding in effectively implementing its rules [40]. More regulatory effort and clearer agreements between the member states are needed here.

We believe that facial recognition is a powerful technique that shows promise to make the world a better place. However, with great power comes great responsibility, which is not always adhered to. Therefore, we need mechanisms that ensure that the technology is used for good purposes, in the interest of humans rather than corporate interest, and that the technology is of high quality (i.e. trustworthy, unbiased and transparent). This not only aligns with the European Commission's ambition to create human-centred AI but will hopefully also increase general trust in AI. However, this requires a clear legal framework. While the population and politicians in the EU are divided on the use of facial recognition technology, effort should be made to avoid that the trust placed in the technology gets undermined.

References

1. van Noorden, R.: The ethical questions that haunt facial-recognition research. Nature **587**, 354–358 (2020)
2. European Commission: Proposal for a regulation of the European Parliament and of the Council laying down harmonised rules on artifcial intelligence (Artifcial Intelligence Act) and amending certain union legislative acts. European Commission, Brussels (2021)
3. Sprokkereef, A., de Hert, P.: Biometrics, privacy and agency. In: Mordini, E., Tzovaras, D. (eds.) The International Library of Ethics, Law and Technology, pp. 81–101. Springer, Dordrecht (2012). https://doi.org/10.1007/978-94-007-3892-8_4
4. Mordini, E., Tzovaras, D.: Second Generation Biometrics: The Ethical, Legal and Social Context, vol. 11. Springer Science & Business Media (2012). https://doi.org/10.1007/978-94-007-3892-8
5. Dantcheva, A., Elia, P., Ross, A.: What else does your biometric data reveal? a survey on soft biometrics. IEEE Trans. Inf. Forensics Secur. **11**(3), 441–467 (2016)
6. Jain, N., Kumar, S., Kumar, A., Shamsolmoali, P., Zareapoor, M.: Hybrid deep neural networks for face emotion recognition. Pattern Recogn. Lett. **115**, 101–106 (2018)
7. Li, Z., Zhang, T., Jing, X., Wang, Y.: Facial expression-based analysis on emotion correlations, hotspots, and potential occurrence of urban crimes. Alex. Eng. J. **60**(1), 1411–1420 (2021)
8. Sim, T., Zhang, L.: Controllable face privacy. In: 2015 11th IEEE International Conference and Workshops on Automatic Face and Gesture Recognition (FG), vol. 04, pp. 1–8 (2015)
9. Chen, X., Duan, Y., Houthooft, R., Schulman, J., Sutskever, I., Abbeel, P.: InfoGAN: interpretable representation learning by information maximizing generative adversarial nets. In: NIPS, pp. 2172–2180 (2016)
10. LeCun, Y., Bengio, Y., Hinton, G.E.: Deep learning. Nature **521**(7553), 436–444 (2015)
11. Sundararajan, K., Woodard, D.L.: Deep learning for biometrics: a survey. ACM Comput. Surv. **51**(3), 1–34 (2018)
12. Fuster, G.G., Peeters, M.A.N.:. Person identification, human rights and ethical principles: rethinking biometrics in the era of artificial intelligence. Technical report, Panel for the Future of Science and Technology (2021)
13. Daly, N., Dickson, A.: Facial surveillance is slowly being trialled around the country (2021)
14. Zalnieriute, M.: How Public Space Surveillance is Eroding Political Protests in Australia (2021)
15. Committee of Ministers: Declaration of the Committee of Ministers on Risks to Fundamental Rights stemming from Digital Tracking and other Surveillance Technologies (2013)

16. Jelinek, A.: Opinion 3/2019 concerning the questions and answers on the interplay between the clinical trials regulation (CTR) and the general data protection regulation (GDPR) (art. 70.1.b)). Technical report, European Data Protection Board (2019)
17. Martini, M.: Regulating algorithms - how to demystify the alchemy of code? In: Ebers, M., Navarro, S.N. (eds.) Algorithms and Law, pp. 100–135. Cambridge University Press, Cambridge (2020)
18. Ebert, A., Spiecker, I.: Döhmann. Der Kommissionsentwurf für eine KI-Verordnung der EU. NVwZ **16**, 1188–1192 (2021)
19. Swedish DPA: Police unlawfully used facial recognition app (2021)
20. Finnish SA: Police reprimanded for illegal processing of personal data with facial recognition software (2021)
21. Dutch DPA issues Formal Warning to a Supermarket for its use of Facial Recognition Technology (2021)
22. Gotink, B.: Slimme camera's herkennen elke carnavalsvierder in Korte Putstraat (2019)
23. Waarlo, N., Verhagen, L.: De stand van gezichtsherkenning in nederland (2020)
24. Montag, L., Mcleod, R., De Mets, L., Gauld, M., Rodger, F., Pełka, M.: The rise and rise of biometric mass surveillance in the EU. Technical report, Edinburgh International Justice Initiative (2021)
25. Dávid, D.: Total Surveillance for 50 Billion - Pintérek's plan has hooked the data protection commissioner (2018)
26. Nagy, M.S.: After terrorists crossed Hungary, surveillance cameras connected through Project Dragonfly (2021)
27. Vass, A.: CCTV: Is It Big Brother or the Eye of Providence?
28. Attila, P.: GDPR Communique to Hungarian Government (2018)
29. Ragazzi, F., Kuskonmaz, E.M., Plájás, I., van de Ven, R., Wagner, B.: Biometric and Behavioural Mass Surveillance in EU Member States. Technical report, The Greens/EFA Group (2021)
30. Halman, L., Luijkx, R., Van Zundert, M.: Atlas of European values. Brill (2005)
31. Mehrabi, N., Morstatter, F., Saxena, N., Lerman, K., Galstyan, A.: A survey on bias and fairness in machine learning. ACM Comput. Surv. (CSUR) **54**(6), 1–35 (2021)
32. Ristani, E., Solera, F., Zou, R., Cucchiara, R., Tomasi, C.: Performance measures and a data set for multi-target, multi-camera tracking. arXiv arXiv:1609.01775 (2016)
33. Zheng, Yu., Capra, L., Wolfson, O., Yang, H.: Urban computing: concepts, methodologies, and applications. ACM Trans. Intell. Syst. Technol. (TIST) **5**(3), 1–55 (2014)
34. Adrian, J., et al.: A glossary for research on human crowd dynamics. Collective Dyn. **4**, 1–13 (2019)
35. Van Rompay, T.J.L., Vonk, D.J., Fransen, M.L.: The eye of the camera: effects of security cameras on prosocial behavior. Environ. Behav. **41**, 60–74 (2009)
36. Pfattheicher, S., Keller, J.: The watching eyes phenomenon: the role of a sense of being seen and public self-awareness. Eur. J. Soc. Psychol. **45**(5), 560–566 (2015)
37. Maras, M.-H.: The social consequences of a mass surveillance measure: what happens when we become the 'others'? Int. J. Law Crime Justice **40**(2), 65–81 (2012)

38. Herring, J., Wall, J.: The nature and significance of the right to bodily integrity. Camb. Law J. **76**(3), 566–588 (2017)
39. Hildebrandt, M., Tielemans, L.: Data protection by design and technology neutral law. Comput. Law Secur. Rev. **29**(5), 509–521 (2013)
40. Zuboff, S.: The Age of Surveillance Capitalism: The Fight for a Human Future at the New Frontier of Power: Barack Obama's Books of 2019. Profile Books (2019)

Emotion Recognition: Benefits and Human Rights in VR Environments

Giulio Barbero[1]([✉])(iD), Richard Albrecht[2], Charlotte Daske[2],
and Marise van Noordenne[1]

[1] Leiden Institute of Advanced Computer Science, Leiden University, Leiden,
The Netherlands
`g.barbero@liacs.leidenuniv.nl`, `m.van.noordenne@umail.leidenuniv.nl`
[2] Department of Information Systems, University of Munster, Munster, Germany
`{r_albr03,c_dask01}@uni-muenster.de`

Abstract. In this paper, we look at the latest developments in the field
of online virtual realities, with particular focus on "metaverses". Within
this field, we aim to describe potential interactions with emotion recognition and manipulation technologies. The article points out how these
technologies can generate new threats to human and individual rights.
Moreover, it covers how already existing uses of these technologies can
become increasingly worrying in the context of metaverses. Finally, we
look at existing legal and technical protections; we focus on their success
and their shortcomings and how they can be improved in order to adapt
to these new technologies. Our research question was: how can emotion
recognition and manipulation technology be used and/or abused in the
context of a metaverse?

1 Introduction

Emotion recognition is being used as a tool for all kinds of applications in both
physical and digital spaces. The fast rise of VR and metaverses can create new
opportunities and synergies for the use of this technology, but questions about
human and individual rights are also being raised. In this paper, the potential
benefits and risks that come with the integration of emotion recognition in virtual reality environments will be discussed. Particular focus will be on online
VR platforms and metaverses, which represent the latest and most promising
development in the field.

Firstly, a general overview of emotion recognition and of virtual reality will
be given. Different types of emotion recognition and the categorization of emotion are elaborated upon. Second, the benefits of emotion recognition will be
explored. Third, the current relation between human rights and the internet will
be discussed. Focusing further, examples of potential human rights violations in
online virtual environments are elaborated upon. Finally, suggestions for limiting
these risks will be discussed.

J. Bayer and C. Grimme (Eds.): AI, Human Rights and Ethics, LNAI 14400, pp. 17–32, 2025.
https://doi.org/10.1007/978-3-031-52082-2_2

1.1 Background

1.1.1 What Is Emotion Recognition?.
Humans display different kinds of emotion on a daily basis which can be understood through facial expressions, gestures, written text, and spoken language. Emotion recognition aims to analyse and interpret sentiments using technology [3,59].

Often, emotion recognition goes along with facial recognition technologies by the detection of the individual's face and expression to classify the emotional state [59]. This type of emotion recognition is called facial emotion recognition.

Other approaches which aim to detect emotions include speech emotion recognition [50], thermal emotion recognition [47], or the most common: text-based emotion recognition [55]. Another approach aims to interpret a human's emotion through electroencephalograms (EEGs), which detect electric activity in one's brain, this is called EEG-based emotion recognition [10].

The aforementioned data is then often interpreted by a pre-trained AI in order to define the corresponding emotion(s). The type of AI varies and provides different insights. For example, long-short term memory (LSTM) neural networks have been deeply investigated in text-based sentiment analysis for their capability to retain "chains" of information (such as words) [4].

The relevance of emotion recognition is increasing, market researchers estimate a compound annual growth rate of 11.3% in the market of emotion detection and recognition during the period from 2020 to 2026 [2].

As of now, emotion recognition can and is being used for different goals ranging from the ability of robots to react to a human's emotional state [57], medical diagnosis support, recruitment of employees [58], in-car emotion detection to increase a driver's safety [11], to many more possible areas of use. According to Gartner, 10% of all personal devices used will have emotion-reading AI capabilities by 2022 [58]. The rapid growth of the field also ensures that new potential uses that are often theorized still face variable limitations in terms of technology, data, and tools availability. This causes the appearance of new platforms (metaverses) and the rapid spread of previously niche technologies (virtual reality) to be extremely promising fields of application.

1.1.2 VR Online.
In the previous years, the rise of virtual reality (VR) has been apparent in the news. The market size of augmented reality (AR) and VR is estimated to rise up to 296.9 billion U.S. dollars until 2024, being 30.7 billion U.S. dollars in 2021 [1].

VR just describes a simulation of three-dimensional objects and environments with user interaction. In a virtual world, multiple users are able to interact remotely in seemingly physical locations, this can be used for work related purposes, communication, or video games. A metaverse goes even beyond that and acts as a large network of virtual worlds [18]. In these virtual environments, users can create and customize their avatars for the purposes of interaction and communication with the environment and other users.

Realism of VR in terms of the users' physical and emotional engagement with the virtual environment is sought after and strived to be enhanced [18]. David et al. describe the possible future of VR and the metaverse and argue that an increase in naturalism in expressions of user avatars leads to a greater perceiving of the reality of the environment [18]. For this purpose, real-time rendering of users through sensors and headphones is an important asset. Sensors within VR headsets allow for multiple input sources for emotion recognition. It is possible to track facial muscle movements, brain monitoring as Marin-Morales et al. [38] did during their study about emotion recognition through the use of wearable sensors, etc. Moreover, the sensors necessary to control the user's perspective can easily provide new information about the user's own awareness. In online settings, this makes it possible to accurately track what the user sees [13] to an entirely new extent.

1.1.3 Categorization of Emotions.
Facial expressions are a great signal for emotion perception, whether for humans or for computers. For the recognition of emotions by a computer, underlying knowledge and data for the classification of emotions is of great importance.

The development of emotion recognition technologies is heavily dependent on the access to databases offering labelled input like images or speech recordings. For facial emotion recognition, public databases offer images of multiple individuals showcasing different expressions and emotions. These images are often labelled according to the corresponding emotion displayed. However, for this, a basic set of emotions needs to be named into which the images can be categorized.

Data-sets available for emotion detection and recognition are for example, the extended Cohn-Kanade Dataset (CK+) [35], Compound Emotion (CE) [19], Binghamton University 3D Facial Expression (BU-3DFE) [61] and the Japanese Female Facial Expressions (JAFFE) [36]. All of these data-sets consider the so called 'six basic emotions' next to a neutral emotion which can be displayed and categorized through facial expressions. The CE-data-set also considers 15 compound emotions [19]. For all of these data-sets the basic emotions are classified as: 'happiness', 'sadness', 'fear', 'anger', 'surprise', and 'disgust'/'contempt'.

This 'Basic Emotion Theory' proposes that humans only have a limited number of emotions. The psychologists Ekman et al. [25] used these for their research on recognition of emotions from facial expressions. However, this list of 'basic emotions' was often shortened or increased by psychologists, including Ekman himself [22,23] who added 'contempt' to the list.

Furthermore, for the recognition of emotions displayed on faces, the 'Facial Action Coding System' (FACS) is often used. This System gives a set of muscle movements which correspond to certain emotions [27]. The FACS mostly used and known today was published by Ekman et al. [24] and describes the muscle movements within the face corresponding to the six basic emotions proposed by Ekman [22,23,25].

However, current psychologists are questioning basic emotions which are universally understood. Especially, the cultural and ethical background of individuals can impact their idea of certain expressions [30]. Furthermore, emotions cannot only be expressed by facial mimicsin fact, people are able to hide and fake emotions through their facial expressions [30]; taking only facial expressions into account for distinguishing emotions can lead to error.

1.2   Relevance

The year 2021 was (also) characterized by a proliferation of sensationalist announcements of new digital places for social interactions: the so-called "meta-verses" [31]. The company Meta (formerly Facebook) presented in October of the same year its version of these online VR social environments [41]. More-over, private companies are not the only ones looking at new virtual environ-ments. Researchers and institutions started developing potential virtual spaces for education [20] and bureaucracy [44]. In order to make the development of the field possible, new tools, technologies, and ideas are necessary. For example, a considerable number of studies focused on users' representation in the virtual reality [40]. This, coupled with new sensors being paired with VR gear [32], allows for reliable and detailed avatars. Others focused on the representation of the virtual space, on new ways to guide users and make them explore the new environment [12, 28].

However, this also translates into more and novel information (often personal) being streamed online, available to a wide audience. It also raises concerns over the effective control users have on the new digital environments. Even though these technologies are extremely recent, at best experimental, others (such as users' emotion recognition and manipulation) are not. The present article aims to start a conversation about potential opportunities and threats that can arise from the interactions between social online virtual realities (such as metaverses) and affective technologies. We focus on emotion recognition and extend to its subsequent applications such as emotion manipulation, psychological and polit-ical profiling and other techniques interacting with the users' emotional status. In the following chapters, we first define the field of research at the intersection of human rights and the Internet. We then speculate over potential abuses but also opportunities arising from the interaction between online VR environments and emotion recognition. We also look at existing regulations, defining their suc-cesses and limitations to influence uses and development of these technologies. Finally, we suggest directions for further explorations; these are aimed to help define changes and additions to the current legislative corpus in order to allow a prosperous development of the field while granting the protection of the users' rights.

2   Human Rights and New Challenges

The interaction between emotion recognition technologies and social metaverses potentially allows for a new degree of access to the human mind. The nature of this data is concerning, especially because its analysis and use are prone to human rights violations. The Universal Declaration of Human Rights (UDHR) by United Nations (1984) consists out of 30 fundamental human rights [43]. With the rise of the internet come both benefits and risks of violation of these human rights. On one hand, the internet allows for almost limitless access to information and opportunities to assemble and express thoughts and opinions on a massive scale, both supporting the right to freedom of thought, conscience, and religion (Article 18) and the right to freedom and expression (Article 19). However, other features, such as anonymity and data collection, can also create opportunities, for example for cyberbullying, cybercrime, discrimination and privacy invasion, posing major risks of violation of the human right to dignity (Article 1), the right to no discrimination of any kind (Article 2), the right to security of person and privacy (Article 3 and 12).

Human rights violations come in many forms and with the rise of the internet the definitions of what is legitimate and what is not became more and more blurred and required serious discussions to be partially redefined. In this spirit, the present paper takes this same perspective towards potential uses of emotion recognition in metaverses. In particular, it focuses on how the technological interactions between the two require new solutions and legal protection to preserve human rights.

3   Human Rights Violations in Virtual Reality

In this section, we go through already existing and theorized applications of emotion recognition in online virtual reality environments (with specific focus on metaverses). The goal is to show that, in the case of technologies and techniques that are already widely used in non-VR environments, virtual reality and its associated tools can greatly enhance emotion recognition effectiveness. In other cases, naturally more limited in current social networks, virtual reality can allow new and more invasive techniques to rise. Moreover, we will see that abuses of emotion recognition in VR environments can generate new threats to human rights.

3.1   Targeted Advertising

The theory behind the developing of advertising messages is deeply rooted in the studies of persuasion and information processing [33]. Even though a detailed description of the overwhelming number of existing models is out of the scope of the present article, most studies individuate a "cognitive awareness" element and an affective component to evaluate advertising effectiveness [8]. The affective component is obviously where emotion recognition and manipulation become central to the discourse, making the message more effective by adapting it to the user's emotional state. Moreover, virtual reality environments can also be powerful tools to influence cognitive awareness [51] providing a certain degree of certainty about what the user is seeing. The natural combination of the two (as described above) poses certain questions about:

- The user's actual freedom in selecting to what advertising content being exposed.
- The use of affective elements being tailored and, therefore, enhanced by emotion recognition.

To summarize, it is important to point out that, in this context, advertising can easily become a relatively advanced form of emotion manipulation, empowered by the potential awareness control of virtual reality.

3.2   Policing and Crime Prevention

The use of emotion recognition in crime prevention has been widely mystified in terms of its current applications. In fact, the field of affective computing is a relatively young one and its applications in crime prevention are currently very limited [5]. This, of course, excluding the more traditional forms of text-based sentiment analyses. However, there are indications that stress levels can be correlated to the individual capacity to commit crime [15]. Current emotion recognition techniques can be effective in the detection of stress levels in various contexts [45]. It is natural to wonder if more precise sensors (such as those available in a VR headset) would further improve this capability; two of the main resources necessary to stress detection are indeed Electro-Dermal Activity (EDA) and Heart Rate Variability (HRV) [5], both "contact" measurements. If these are difficult to retrieve in current social networking contexts, they can become easier to record in VR environments (due to VR headsets). Moreover, the naturally improved awareness control of VR online environments that we described above can arguably be effectively used to design stress tests aimed at crime intention detection. This poses new risks in terms of privacy and human rights:

- The collection of data previously challenging to gather such as EDA and HRV can become normality and officially justified in terms of security.
- "Crime capacity tests" can invade the users' personal space and be considered more reliable than previously.
- As a result of biased data sets, individuals can be monitored unfairly, further exacerbating racial biases.

In conclusion, if current crime prevention and policing techniques seem unable to make a massive use of emotion recognition, online VR environments such as metaverse based social networks can quickly change this due to more complete measurements and awareness control.

3.3   Dissidents Repression

Emotion recognition can have some application in the repression of political dissidents [42]. However, it is important to point out that the use of facial emotion recognition in this regard has been mainly theorized and real-world applications have been sparse. On the other hand, text-based sentiment analysis has proven promising in predicting political shifts in dissidents organizations [53]. As for the case of crime prevention, we can currently only infer possible interactions between emotion recognition and social metaverses. In this direction, it is important to notice that tests guided by awareness control techniques can arguably be successfully applied in this field. Users can be exposed to testing images (not necessarily directly related to political content) to determine their political tendencies [56]. As already remarked above, VR headset can easily accommodate measuring tools already used in experimental settings [34] in order to infer this type of information [32]. This is a field of particular concern for multiple reasons:

- Using emotion recognition for this purpose represents an obvious threat to civil and political rights (Article 2 and 18 of the UDHR [43]), strongly impairing democratic discourse.
- Perhaps more worryingly, using emotion recognition in a VR headset as a measuring tool would provide access to personal non-verbal cues (and, by extension, emotions). This type of information, when applied to dissident repression, opens to a clear violation of the right to privacy as formulated in Article 12 of the UDHR [43].

4 Benefits of Emotion Recognition

The previous chapter focused on possible threats rising from the interaction between emotion recognition and metaverses. On the other hand, emotion recognition can be used in a variety of fields with different benefits [60]. In this chapter, we focus on these and expand on the opportunities that can be brought forward in the same intersection (emotion recognition and online VR).

4.1 Healthcare

Emotion recognition has the potential to e.g. be used to assist medical personnel in determining whether a patient requires medical attention [7]. For this purpose, virtual environments could be set up to enable communication between patients and doctors. In these environments, emotion recognition software could be used to aid the doctor's decision making process on how to treat the patient. This can also be helpful for psychological treatments, since a psychologist could be supported in her assessment of the patients' emotional state [7,39]. Emotion recognition could also be used in helping disabled individuals. In an attempt to use it to aid children with Asperger syndrome, a system can be developed that allows them to read other people's emotions, to react and to learn them [26]. In conclusion, emotion recognition in metaverses opens new exciting opportunities in the field of e-health, making treatments more personal, appropriate and providing previously hidden information to healthcare providers.

4.2 Tailored Advertising

Through emotion recognition, advertisements can be tailored towards a specific user and recommendations incorporating emotional feedback could be made [39]. Especially in virtual environments, it is possible to show advertisements that interest users. The incorporation of real time emotional feedback provides the opportunity to be used to present more relevant advertisements to the user. This might be beneficial to advertisers because it allows them to show their products to potentially interested customers. Companies can not only optimize advertisements using emotional recognition software, but can also optimize their products by analysing users' interactions with a product. The emotional feedback received can be used to identify problems that users experience while using a product. In this regard, we must define the difference between "tailored advertising", the opportunity to present products that effectively excite users' interest and "targeted advertising" (see chapter "Human Rights Violation in Virtual Reality"), which we interpret as the dangerous exploitation of users' emotional states to sell products. While the first uses customers' previous interactions to learn individual preferences (therefore tailored around the individual), the second constantly monitors users' feelings in order to present certain products at the most profitable moment (therefore targeting users in a certain emotional state).

4.3 Engagement and Entertainment

Emotion recognition software can also be used to enhance the engagement of a user with the virtual world [37], and thereby allow a user to have better experiences. Software developers could use the technology to receive valuable feedback about products or the virtual environment itself. This would allow for a more precise and more profound knowledge about users' engagement, which information is especially valuable for the entertainment industry. This can be beneficial to all parties since the users will receive enhanced entertainment service and content creators have the opportunity to optimize their content. Generally, this type of information can be used to develop new and smart tools to personalize users' experiences. For example, emotion recognition software can be used to play music that fits to the mood of a user [62]. Furthermore, similar techniques can be beneficial for educational purposes in virtual learning environments. Emotion recognition can be used to adjust learning material and to observe students so as to keep them engaged with the topic. Therefore, emotion recognition software can lead to better learning results [54]. This could also help teachers to adjust their teaching methods using deep emotional feedback that would not be available in the real world.

Furthermore, there are approaches to create virtual reality environments to enable design thinking. For this purpose, emotional recognition software can be used to optimize the virtual environment so that it enables the creative thinking process of users.

5   Existing Regulations and Future Suggestions

The two previous chapters defined both threats and opportunities arising from emotion recognition and metaverse technologies. Generally, most of the uses (and abuses) we mentioned require a certain level of tracking and personal information to be shared. In this regard, emotion recognition in virtual environments can bring benefits to both users and content creators. However, there are also risks and legal concerns involved. To mitigate these risks, regulations can be put into place, similar to what the European Union proposed for regulating AI in general [16]. These rules are not specialized for emotion recognition in virtual environments. In this section, current regulations, as well as proposals made by the European Union, are discussed.

5.1   Legal

As for legal regulations, there already exist some within the General Data Protection Regulation (GDPR) [17]. However, aspects of the GDPR are criticized, e.g. the right to be forgotten because they are partially infeasible to implement [49]. Another issue of the GDPR is that in certain cases, it hinders research [48], therefore potential benefits of data processing are lost. As mentioned previously, the GDPR is mainly focused on data and privacy protection and does not specialize on regulating Artificial Intelligence or Virtual Realities. Since these technologies are innately different from previous data processing algorithms, new regulations that specialize on Artificial Intelligence and virtual reality need to be put into place.

To regulate AI, the European Commission set forth a proposal that aims to "turn Europe into the global hub for trustworthy Artificial Intelligence" [16]. It seeks to establish trust towards AI by regulating the usage of AI. To achieve that, the European Commission classifies four risk categories: Unacceptable risk, High-risk, Limited risk and Minimal risk [16]. Each AI system has to be classified into one of these categories and is then supposed to be regulated according to their risk class. These regulations also regulate the data that AI systems might use for training and analysis.

The proposal for the regulations is also criticized [21]. The topic of emotion recognition via Artificial Intelligence is rather specific. Therefore, more specific measures might be necessary and no general regulations for all AI systems can cover all aspects [21]. Therefore, the proposal needs improving before it can be translated into regulations to minimize the dangers of AI while obtaining its potential benefits.

Within the proposal, regulations for biometric identification are made. For example, the use of remote biometric identification falls into the high-risk category and is therefore subject to strict requirements [16]. Emotion recognition software in virtual environments falls into the category of remote biometric identification. Therefore, it might be classified as a high-risk application. However, emotion recognition in virtual environments has different potential uses compared to its real-world applications, therefore, different and stricter regulations should be applied to protect users. In this regard, in the chapters above, we point out how emotion recognition in metaverses has the potential to enhance emotion manipulation techniques and their effects. Therefore, it could be classified into the unacceptable risk category [16].

Since the proposal is not focused on virtual environments, additional regulations that include these should be proposed and enforced. In order to regulate these new realities, it is possible to apply existing categories when possible. However, it is inevitably necessary to define and create new legal definitions in order to cover the intrinsic novelty of Metaverses. The proposed Metaverse could e.g. be considered as a digital/physical public space that needs to adhere to all the existing regulations. In order to do so, it will be necessary to develop for it regulations similar in effect to the GDPR, in which we separate data gathered through emotion recognition techniques from user profiling methods. This regulation should clearly and effectively inform users of the type of data they are streaming on VR online platforms. Additionally, it should be ensured, that users benefit from the use of their data [46]. Moreover, further regulation should specifically make the tracking and sharing of physical parameters (for example facial expressions but also heartbeat) optional or, considering the aforementioned applications of this information, effectively anonymous. This would empower users with the possibility to choose freely and without repercussions on their virtual experience. For this case, the Artificial Intelligence Video Interview Act [6] could be applied to the Metaverse. However, it is important to remark that if certain virtual environments used for research purposes are obliged to follow the same rules, research in this field could be hindered. This can be tackled by specifically building exceptions, as already done and stated in the goals of the aforementioned European Commission's proposal about AI [16].

To enforce regulations, obligations are put upon providers of such services [16]. Similar obligations can be put on the providers of virtual environments as well as the providers of the software for that environment. In order to check the providers, transparency requirements must be met in the storage and presentation of the gathered data. Controllers need to be able to check whether the data gathered through emotion recognition techniques is used in an ethical manner, respecting the individual's rights.

In parallel, it will be necessary to create institutions to ensure human oversight of virtual worlds that use emotion recognition software to identify when it is used in a manipulative way [21]. These institutions should monitor both the technical side of virtual environments (with specific expertise and access to information about how data is streamed, gathered, processed and possibly anonymized and how AI possibly uses it) and also the social structures which would spontaneously emerge within the environment. In this sense, it is essential to be aware of how HR violations might also happen inside these structures, carried out by users themselves. For example, a case of sexual assault already took place in the virtual reality platform Horizon Worlds [29]. Meta reacted by updating the personal space tools available to its users. However, this approach does not suffice with larger traffic and therefore, institutions with legal knowledge and power are necessary to intervene and deter [9,52].

Responsible institutions need to pay continued attention [14] towards new developments in Metaverses to be able to update their regulations efficiently in order to tackle new types of abuse. In order to effectively carry out their tasks, the institutions should be granted special access to the providers platforms. However, it is important to remark that the regulation to build such powerful institutions must pay attention to prevent possible abuses from these institutions themselves. After the creation of new regulations, the intended effects and actual outcomes should also be evaluated and adjustments have to be made accordingly [46].

Applying these regulations and structures, it would be possible to enjoy the benefits of emotion recognition that have been mentioned in this article, while granting users a safe environment where their individual freedoms are respected. The mentioned regulations are also supposed to protect and enhance the development of the research in the field, which is also the goal of the European Commission's proposal [16]. The intended trust in technology is also expected to motivate further investments into it. Additionally, creating unified rules for the European Union, which creates a large market for providers of emotion recognition software for virtual environments. All regulations that are proposed have to make sure that users are assured that their human rights are not violated. However, it cannot come at the cost of innovation in that field, since it can also provide a multitude of benefits (see Sect. 4). Therefore, currently existing regulations and proposals on data privacy like the GDPR need to be expanded in order to suit the intrinsic novelty of virtual environments.

6   Discussion

Virtual reality technology embedded with online social interactions can undoubtedly bring new opportunities and exciting changes to the way we experience the digital world. Metaverses promise to generate customized and creative spaces for everyone to "live". Emotion recognition technology will also profit from these new opportunities, granting people new power to communicate and express themselves. However, the combination of metaverses and emotion recognition is a particularly worrisome one, since online VR renders service providers opportunities to control their users' data and what content they are exposed to. Moreover, VR gear can pack new sets of sensors able to leak personal information that was never available before. In turn, these developments can be exploited to make emotion recognition and manipulation techniques highly dangerous to individual freedoms and human rights.

In this regard, the new data available grants both public and private stakeholders new access to the users' personal opinions, plans, and even thoughts while the awareness control available through VR environments will bring novel and more effective methods to manipulate individuals' and groups' behaviour. While our current legal tools are not completely prepared to deal with the new complexity arising from the interactions of these (and other subsequent) technologies, this article aims to start a conversation in advance, while monitoring the development of metaverses. The goal is to update the protections available in time and to be able to communicate risks and opportunities in a complete and informed way.

6.1 Summary

The interaction between online virtual reality spaces (metaverses) and emotion recognition technology can potentially:

- empower already existing techniques giving more in-depth access to individuals' opinions and thoughts
- generate new forms of emotion and behaviour manipulation techniques, selecting and controlling what users experience in the virtual world
- provide access to new personal and private information previously unavailable

All these new possibilities will obviously represent major threats to both individual freedoms and human rights. In many cases, service providers could potentially monitor, profile and manipulate users while de facto controlling to what content they are exposed. Even though the infrastructure to tackle these issues exists, it should be updated in order to cover these new future realities. The current article strongly calls for a more active conversation in the field in order to start the process to extend the scope of the current legal protection to include metaverses and online VR in general.

6.2 Future Research

The present article is centred on a very new technology which, on the one hand, is still in its infancy. On the other, the massive investments in resources made by both public and private stakeholders needs to be taken into account; it is likely that VR online "worlds" or metaverses will develop rapidly in the coming years. Future research should start to collect information about how developers are currently building these environments and with which features. These features can then be used to anticipate more precisely possible misuses and human rights threats. Future studies should also consider global social and political developments, to spot possible abuses that could potentially arise from these technologies, so that early warnings of emerging threats can be given. We also recommend closely following the developments in the field of avatar creation; the way we portray and represent ourselves can tell a lot about what type of (personal) information we broadcast to the world. Finally, future research should focus on new solutions to update and refine our current tools for legal protection and perhaps propose completely novel ones.

References

1. Augmented reality (AR) and virtual reality (VR) market size worldwide from 2016 to 2024 (in billion U.S. dollars) [graph] (2022). https://www.statista.com/statistics/591181/global-augmented-virtual-reality-market-size/
2. Emotion detection and recognition market size, share and global market forecast to 2026 — marketsandmarkets (2021). https://www.marketsandmarkets.com/Market-Reports/emotion-detection-recognition-market-23376176.html?
3. Emotion recognition: Introduction to emotion reading technology (2021). https://recfaces.com/articles/emotion-recognition
4. Alhagry, S., Aly, A., El-Khoribi, R.A.: Emotion recognition based on EEG using LSTM recurrent neural network. Int. J. Adv. Comp. Sci. Appl. **8**(10), 355 (2017)
5. Anderez, D.O., Kanjo, E., Amnwar, A., Johnson, S., Lucy, D.: The rise of technology in crime prevention: opportunities, challenges and practitioners perspectives, pp. 1–19 (2021). http://arxiv.org/abs/2102.04204
6. Assembly, I.G.: Artificial intelligence video interview act. Retrieved January **13**, 2021 (2020)
7. Ayata, D., Yaslan, Y., Kamasak, M.E.: Emotion recognition from multimodal physiological signals for emotion aware healthcare systems. J. Med. Biol. Eng. **40**(2), 149–157 (2020)
8. Barry, T.E.: The development of the hierarchy of effects: an historical perspective. Curr. Iss. Res. Advertising **10**(1–2), 251–295 (1987)
9. Basu, T.: The metaverse has a groping problem already (2021). https://www.technologyreview.com/2021/12/16/1042516/the-metaverse-has-a-groping-problem
10. Bos, D.O., et al.: EEG-based emotion recognition. Influence Vis. Auditory Stimuli **56**(3), 1–17 (2006)
11. Bosch, E., et al.: Emotional garage: a workshop on in-car emotion recognition and regulation. In: Adjunct Proceedings of the 10th International Conference on Automotive User Interfaces and Interactive Vehicular Applications, pp. 44–49 (2018)
12. Soo Choi, H., Heon Kim, S.: A content service deployment plan for metaverse museum exhibitions-centering on the combination of beacons and HMDs. Int. J. Inf. Manage. **37**(1), 1519–1527 (2017). https://doi.org/10.1016/j.ijinfomgt.2016.04.017
13. Clay, V., König, P., König, S.U.: Eye tracking in virtual reality. J. Eye Mov. Res. **12**(1) (2019)
14. (CNIL), F.D.P.A.: How can humans keep the upper hand? report on the ethical matters raised by AI algorithms (2017)
15. Colvin, M., Cullen, F.T., Ven, T.V.: Coercion, social support, and crime: an emerging theoretical consensus. Criminology **40**(1), 19–42 (2002)
16. Commision, E.: Europe fit for the digital age: Commission proposes new rules and actions for excellence and trust in artificial intelligence. Geneva, Switzerland, Europan Commision (2021)
17. Commission, E.U.: 2018 reform of EU data protection rules. https://ec.europa.eu/commission/sites/beta-political/files/data-protection-factsheet-changes_en.pdf
18. Dionisio, J.D.N., Iii, W.G.B., Gilbert, R.: 3D virtual worlds and the metaverse: current status and future possibilities. ACM Comput. Sur. (CSUR) **45**(3), 1–38 (2013)
19. Du, S., Tao, Y., Martinez, A.M.: Compound facial expressions of emotion. Proc. Natl. Acad. Sci. **111**(15), E1454–E1462 (2014)

20. Duan, H., Li, J., Fan, S., Lin, Z., Wu, X., Cai, W.: Metaverse for social good: a university campus prototype. In: MM 2021 - Proceedings of the 29th ACM International Conference on Multimedia, pp. 153–161 (2021). https://doi.org/10.1145/3474085.3479238
21. Ebers, M., Hoch, V.R., Rosenkranz, F., Ruschemeier, H., Steinrötter, B.: The European commission's proposal for an artificial intelligence act-a critical assessment by members of the robotics and AI law society (rails). J **4**(4), 589–603 (2021)
22. Ekman, P.: Basic emotions. Handbook of Cognition and Emotion **98**(45–60), 16 (1999)
23. Ekman, P., Cordaro, D.: What is meant by calling emotions basic. Emot. Rev. **3**(4), 364–370 (2011)
24. Ekman, P., Friesen, W.V.: Facial action coding system. Environ. Psychol. Nonverbal Behav. (1978)
25. Ekman, P., Sorenson, E.R., Friesen, W.V.: Pan-cultural elements in facial displays of emotion. Science **164**(3875), 86–88 (1969)
26. El Kaliouby, R., Robinson, P.: The emotional hearing aid: an assistive tool for children with asperger syndrome. Univ. Access Inf. Soc. **4**(2), 121–134 (2005)
27. Farnsworth, B.: Facial action coding system (facs) - a visual guidebook (2019). https://imotions.com/blog/facial-action-coding-system/#emotions-action-units
28. Gómez-diago, G.: Physical space and the flow of communication : an explanatory approach to the (November 2010) (2015). https://doi.org/10.1386/mvcr.1.1.51
29. Heath, A.: Meta opens up access to its VR social platform horizon worlds (2021). https://www.theverge.com/2021/12/9/22825139/meta-horizon-worlds-access-open-metaverse
30. Heaven, D.: Why faces don't always tell the truth about feelings. Nature **578**, 502–505 (2020)
31. Kraus, S., Kanbach, D.K., Krysta, P.M., Steinhoff, M.M., Tomini, N.: Facebook and the creation of the metaverse: radical business model innovation or incremental transformation? Int. J. Entrepreneurial Behav. Res. **28**, 52–77 (2022)
32. Kritikos, J., Tzannetos, G., Zoitaki, C., Poulopoulou, S., Koutsouris, P.D.: Anxiety detection from electrodermal activity sensor with movement interaction during virtual reality simulation. In: International IEEE/EMBS Conference on Neural Engineering, NER, pp. 571–576 (2019). https://doi.org/10.1109/NER.2019.8717170
33. Leung, X.Y., Lyu, J., Bai, B.: A fad or the future? examining the effectiveness of virtual reality advertising in the hotel industry. Int. J. Hospitality Manage. **88**, 102391 (2020). https://doi.org/10.1016/j.ijhm.2019.102391
34. Liu, Y., Du, S.: Psychological stress level detection based on electrodermal activity. Behav. Brain Res. **341**, 50–53 (2018)
35. Lucey, P., Cohn, J.F., Kanade, T., Saragih, J., Ambadar, Z., Matthews, I.: The extended cohn-kanade dataset (ck+): a complete dataset for action unit and emotion-specified expression. In: 2010 IEEE Computer Society Conference on Computer Vision and Pattern Recognition-Workshops, pp. 94–101. IEEE (2010)
36. Lyons, M., Akamatsu, S., Kamachi, M., Gyoba, J.: Coding facial expressions with gabor wavelets. In: Proceedings Third IEEE International Conference on Automatic Face and Gesture Recognition, pp. 200–205. IEEE (1998)
37. Malcangi, M.: Smart recognition and synthesis of emotional speech for embedded systems with natural user interfaces. In: The 2011 International Joint Conference on Neural Networks, pp. 867–871. IEEE (2011)
38. Marín-Morales, J., et al.: Affective computing in virtual reality: emotion recognition from brain and heartbeat dynamics using wearable sensors. Sci. Rep. **8**(1), 1–15 (2018)

39. McDuff, D., el Kaliouby, R., Senechal, T., Amr, M., Cohn, J.F., Picard, R.: Affectiva-mit facial expression dataset (am-fed): naturalistic and spontaneous facial expressions collected "in-the-wild". In: 2013 IEEE Conference on Computer Vision and Pattern Recognition Workshops, pp. 881–888 (2013). https://doi.org/10.1109/CVPRW.2013.130

40. Merola, N., Peña, J.: The effects of avatar appearance in virtual worlds. J. Virtual Worlds Res. **2**(5) (1970). https://doi.org/10.4101/jvwr.v2i5.843

41. Meta: The facebook company is now meta — meta (2021). https://about.fb.com/news/2021/10/facebook-company-is-now-meta/

42. Mohammad, S.M.: Ethics sheet for automatic emotion recognition and sentiment analysis (September) (2021). http://arxiv.org/abs/2109.08256

43. Nations, U.: Universal declaration of human rights (1948). https://www.un.org/en/about-us/universal-declaration-of-human-rights

44. Ning, H., Wang, H., Lin, Y., Wang, W., Dhelim, S.: A survey on metaverse: the state-of-the-art, technologies, applications, and challenges 1 introduction 2 recent advances of the metaverse. IEEE Internet Things J. **10**, 14671–14688 (2023)

45. Ollander, S., Godin, C., Charbonnier, S., Campagne, A.: Feature and sensor selection for detection of driver stress. In: PhyCS, pp. 115–122 (2016)

46. Ong, D.C.: An ethical framework for guiding the development of affectively-aware artificial intelligence. In: 2021 9th International Conference on Affective Computing and Intelligent Interaction (ACII), pp. 1–8. IEEE (2021)

47. Ordun, C., Raff, E., Purushotham, S.: The use of AI for thermal emotion recognition: a review of problems and limitations in standard design and data. arXiv preprint arXiv:2009.10589 (2020)

48. Peloquin, D., DiMaio, M., Bierer, B., Barnes, M.: Disruptive and avoidable: GDPR challenges to secondary research uses of data. Eur. J. Hum. Genet. **28**(6), 697–705 (2020)

49. Politou, E., Alepis, E., Patsakis, C.: Forgetting personal data and revoking consent under the GDPR: challenges and proposed solutions. J. Cybersecurity **4**(1), tyy001 (2018)

50. Ramakrishnan, S., El Emary, I.M.M.: Speech emotion recognition approaches in human computer interaction. Telecommun. Syst. **52**(3), 1467–1478 (2013). https://doi.org/10.1007/s11235-011-9624-z

51. Schwienhorst, K.: Why virtual, why environments? implementing virtual reality concepts in computer-assisted language learning. Simul. Gaming **33**(2), 196–209 (2002)

52. Sheera Frenkel, K.B.: The metaverse's dark side: here come harassment and assaults (2021). https://www.nytimes.com/2021/12/30/technology/metaverse-harassment-assaults.html

53. Shellman, S.M., Levey, B.P., Young, J.K.: Shifting sands: explaining and predicting phase shifts by dissident organizations. J. Peace Res. **50**(3), 319–336 (2013). https://doi.org/10.1177/0022343312474013

54. Shen, L., Wang, M., Shen, R.: Affective e-learning: using "emotional" data to improve learning in pervasive learning environment. J. Educ. Technol. Soc. **12**(2), 176–189 (2009)

55. Shivhare, S.N., Khethawat, S.: Emotion detection from text. arXiv preprint arXiv:1205.4944 (2012)

56. Smith, K.B., Oxley, D., Hibbing, M.V., Alford, J.R., Hibbing, J.R.: Disgust sensitivity and the neurophysiology of left-right political orientations. PLoS ONE **6**(10), e25552 (2011). https://doi.org/10.1371/journal.pone.0025552

57. Spezialetti, M., Placidi, G., Rossi, S.: Emotion recognition for human-robot inter-action: recent advances and future perspectives. Front. Robot. AI **7**, 532279 (2020)
58. Susan Moore: 13 surprising uses for emotion AI technology (2018). https://www.gartner.com/smarterwithgartner/13-surprising-uses-for-emotion-ai-technology
59. Vemou, K., Horvath, A., Zerdick, T.: Facial emotion recognition (2021). https://edps.europa.eu/data-protection/our-work/publications/techdispatch/techdispatch-12021-facial-emotion-recognition_en
60. Vinola, C., Vimaladevi, K.: A survey on human emotion recognition approaches, databases and applications. ELCVIA Electron. Lett. Comput. Vis. Image Anal. **14**(2), 24–44 (2015)
61. Yin, L., Wei, X., Sun, Y., Wang, J., Rosato, M.J.: A 3D facial expression database for facial behavior research. In: 7th International Conference on Automatic Face and Gesture Recognition (FGR06), pp. 211–216. IEEE (2006)
62. van der Zwaag, M.D., Janssen, J.H., Westerink, J.H.: Directing physiology and mood through music: validation of an affective music player. IEEE Trans. Affect. Comput. **4**(1), 57–68 (2012)

Facial Recognition Technologies (FRT) in Public Spaces and Their Impact on Individual and Collective Behavior

Fareeha Saeed[✉] and Lennart Schmidt

Westfälische Wilhelms-Universität Münster, Münster, Germany
Fareehasaeed1996@gmail.com, l_schm80@uni-muenster.de

Abstract. Facial Recognition Technologies (FRT) have the potential to be prevalent in society. Being constantly exposed to such technology could have an impact on the individual and collective behaviour of individuals and society as a whole. In this paper, we try to explore the terrain of the history of surveillance and its modern form in places like China to set the context and introduce current technologies. Later, we explore the possible ways the society could change if they were under constant observation by the FRT and what dangers can arise over time.

1 Introduction

Surveillance in public spaces with CCTV cameras is not a new concept. Most times, the tapes are recorded and retrieved when in need. CCTV systems had been in place as mere data collectors for a long while, but those days are over with the new advances in modern technology. Technology now enables us to tread one step ahead. Although CCTV surveillance is still a kind of well-known public space monitoring, the development of technology in the surveillance field now makes it possible to actively monitor the subjects as well as passively record them. Artificial intelligence algorithms and CCTV cameras are now able to work together in order to create intelligent observers who can recognize subjects and track their whereabouts. This idea does not come from utopia, but such technologies are implemented in a few places worldwide. One of the prime examples is Xinjiang, China. The idea of installing such technology in public spaces invokes several questions. One of such questions concerns how 'being watched' impacts the subject's behaviour. In this paper, we explore the terrain of public space surveillance with facial recognition technology from the lens of behavioural psychology and try to find out in what ways FRT surveillance is likely to impact the behaviour of citizens on an individual level and how does that in turn impacts society's collective behaviour as a whole.

J. Bayer and C. Grimme (Eds.): AI, Human Rights and Ethics, LNAI 14400, pp. 33–46, 2025.
https://doi.org/10.1007/978-3-031-52082-2_3

2   Theoretical and Ethical Context

2.1   Jeremy Bentham's Panopticon

An all-watching, perfect surveillance mechanism was presented by Jeremy Bentham in the form of the Panopticon in the 18th century [3]. The idea behind the Panopticon was to have a surveillance system in a place where the subjects would discipline themselves. Bentham himself called it "a mill for grinding rogues honest" [4]. It was an architectural attempt to perfect and rationalize a surveillance mechanism in which the subjects know they are being watched all the time, but cannot see the watchman. Bentham regarded the Panopticon as humane and efficient, and regarded it as "a new mode of obtaining power of mind over mind." The Panopticon, as envisioned by Bentham, was heavily debated and criticized. If an individual is entirely isolated and visible, then power functions automatically, says philosopher Michel Foucault [13]. 'The surveillance is permanent in its effects, even if it is discontinuous in its action. [...] the perfection of power should tend to render its actual exercise unnecessary' [13]. Bentham's Panopticon makes for a good prison setting. Surprisingly, this prison lacks bars, chains, or heavy locks. All that is needed is for the mind to exercise its power over itself. Installing facial recognition surveillance in public spaces essentially means converting the state into a big Panopticon where every citizen is a subject and thus knows that he is being watched at all times. In essence, the subjects in Bentham's Panopticon are never able actually to see the observer, but they are fully aware of the observer's presence. FRT surveillance carries the power to turn societies into practical implementations of Bentham's Panopticon. Every citizen knows that he is being watched at all times, and thus becomes the observer of his own actions.

2.2   Surveillance in an Ethical Context

To discuss surveillance in the context of ethics, it is necessary to look into the question of why it is put in place. The purpose of the surveillance in a public space could decide whether it is ethical in a particular context. One of the examples is the usage in retail stores for their and their customers' advantage. Another and perhaps more apparent use of surveillance is for security. When exploring surveillance in an ethical context, the idea of consent must be considered. There are two main approaches to justifying surveillance's ethical aspects: the consequentialist approach and the deontological approach [21]. The former regards the consequences of an action as the basis for judging if the action is right or wrong. However, the latter judges the morality of an action based on whether the action in itself is right or wrong. In the case of public space FRT surveillance, a consequentialist approach would support it. If the surveillance is conducted at a state level, it is virtually impossible to gain the consent of those involved, even if it is improbable that all the citizens would consent to it. In this case, the consequentialist approach would argue that the greater good and safety of the citizens justify the surveillance of everyone. However, the deontologist approach would not agree, and it would argue that everyone has equal rights and that if a small portion of the subjects does not consent, then it retains the right not to be spied on. When putting surveillance systems in practice, both approaches to the justification of surveillance need to be considered.

Regarding the ethics of surveillance, another authoritative consideration is Article 8 of the European Convention on Human Rights. This concerns 'The right to respect for privacy, and family life with the following two points [12]:

1. "Everyone has the right to respect for his private and family life, his home, and his correspondence."
2. "There shall be no interference by a public authority with the exercise of this right except such as is in accordance with the law and is necessary for a democratic society in the interests of national security, public safety or the economic well-being of the country, for the prevention of disorder or crime, for the protection of health or morals, or the protection of the rights and freedoms of others."

Article 8 §1 talks about "private life" The definition of it could be extremely broad or highly narrow; therefore, it is crucial to decide the bounds of what should be considered a part of a person's private life. The European Court of Human Rights argues that the purpose of Article 8 is that it protects the public authority to a person's private, family life, home, and correspondence [14].

Further, Article 8 §2 entails the aims that justify the interference and infringement on the rights that are protected by Article 8 §1: "in the interests of national security, public safety or the economic well-being of the country, for the prevention of disorder or crime, for the protection of health or morals, or the protection of the rights and freedoms of others". With regard to the legitimation of FRT in public spaces under the context of Article 8 §2, it is crucial to reflect on the following:

1. To what ends does FRT surveillance provide means?
2. Whether the risk is to national security, public safety, or the country's economic well-being is high enough so that Article 8 §2 applies?

2.3 Current Techniques, Possibilities, and Limitations

The term facial recognition describes the process of identifying facial features of individuals and deriving information from them. Although this concept is often times associated with modern and computer-aided technical artefacts, it originated from our own human experiences of how we detect and identify human faces as well as assign them to other familiar faces from our own point of view. Psychologists Bruce and Young [9] developed a theoretical model for understanding how humans recognize faces and which information can be derived from observed faces for further processing. They suggested specific types of information such as structural, visually derived semantic and identity-specific semantic, which are of great importance for differentiating human faces from ordinary objects and matching faces to familiar persons in everyday life. Other more technical fields of studies, such as computer science and data science, have produced a growing body of literature through investigating and developing algorithms, techniques, and systems to automate the process of facial recognition [32]. Throughout this paper, we use the broader term facial recognition technology (FRT) to refer to technical artefacts of such kind [15]. Implementing advanced algorithms for FRT needs to consider many aspects to fulfil the requirements of identifying faces under various conditions in a robust way [27]: before analysing facial features themselves, the face within the picture must be detected and set apart from other visual shapes. Considering video material, it is detecting necessary to track the face over the frames and adapt to the changing visual conditions. Algorithms must be capable to detect faces from different distances, whereby for high-distance objects the quality of images can suffer, and fewer details are available for analysis. To fulfil these requirements, algorithms have to take into account the perspective, illumination, and angle of view as well as specific human features such as skin colour, age, hairstyle, facial accessory, or masks of which some can be dynamic but all influence the overall appearance of individuals.

One famous algorithm adapted by Turk and Pentland [26] is called *Eigenfaces*, which utilizes principal component analysis (PCA). Under controlled circumstances, it is capable of locating and tracking individuals' heads and recognizing persons by comparing specific facial features of them to already known individuals, all of it in near-real-time. It makes use of a 2-D characteristic view and creates so-called 'Eigenfaces', which are vectorized representations of the facial features of the faces in the knowledge database (training set) with values, that account for the most variance in the training set. It is similar to how humans focus on specific parts of the face to identify individuals such as eyes, nose, mouth, or ears, but the algorithm is more sophisticated and locates its own set of features, with which it can achieve the best identification scores. Each face can be approximated as a linear combination of the Eigenfaces as a weight vector. For new spotted faces, the image can again be transformed to a vector, and weights are calculated for the representation through Eigenfaces. After that, distances to weight vectors of all previous processed and labelled faces determine an allocation to an existing or the creation of some new individual in the database. Location and tracking of people over time in videos are reached by simple motion detection and tracking algorithms taking advantage of frame differencing, motion bulbs, and potential head-to-body proportions of detected moving objects. There exist other appearance-based methods which make use of artificial intelligence and machine learning techniques such as Neural Networks, Support Vector Machines, or Naive Bayes classifier [29]. Others take predefined templates into account to analyse facial similarities based on the whole face or on separate facial features pattern-wise [10]. Finally, some authors even compared human and machine-based recognition approaches in general and tried to standardize benchmark processes across disciplines and to extract synergism for future research [11].

The above-mentioned requirements of robust FRT algorithms outline a first overview of possible challenges and their limitations. Notable among others are the optics and decreasing light intensity for increasing distances, exposure time and blur in moving pictures, image resolution in combination with illumination in general and inconsistent pose and facial expressions of individuals [27]. Of course, there must exist some form of database of existing facial images such that individuals can be assigned to their corresponding identity. Building, forming, or extracting these databases can be a lengthy process, and especially the gradual changes of external facial features over time can be problematic here.

If we transfer those challenges to the surveillance topic of this paper, many aspects can be neutralized. First, the more surveillance cameras are set up, the more angles, perspectives, and situations are captured, which significantly increases the chances of catching usable facial excerpts. Furthermore, establishing such cameras comes with the possibility to determine the exact placement and influence the illumination through artificial light or natural bright places. Certain positions facilitate capturing frontal images of the face, for example escalators, elevators, entrance-only/exit-only doors, service counters, security checks. Especially the latter can be exploited so that people are required to look into these FRT cameras because it is an essential part to passing the check itself. Camera resolution quality for low to middle-budget devices is continuously improving over the last 15 years, which is particularly noticeable through modern smartphone technology. Similar improvements can be expected for surveillance cameras, which advance their image quality, reduce their size, lower their cost and thereby facilitate a denser, but less prominent coverage of populated areas in the public space.

In contrast, surveillance cameras alone without FRT can just be used in primitive ways. They record all activities inside their perspective without identifying people ad-hoc for further processing. Authorized persons can utilize this kind of record to investigate incidents shown there or, if watched in real-time, they can interfere. One could describe it as a more passive form of surveillance with active human interactions if needed, while the extension of FCT creates a system of continuous monitoring with active forms of surveillance with rather passive human interaction. To summarize the findings of FRT's range of possibilities, we can assert that with modern technology and sufficient camera coverage, surveillance systems of such kind are indeed capable of identifying human beings by their face, tracking them over single viewpoints, recognizing them across multiple locations and thereby create movement and behaviour profiles.

2.4   Public Acceptance of FRT in Public Spaces

Even before asking of how mass surveillance with FRT in public spaces influences the behaviour of people under such observations, we need to consider the attitude towards the implementation of FRT systems. Negative views and missing acceptance can strongly influence how people handle situations with the presence of FRT and lead to conscious and unconscious behaviour changes. At the same time, strong advocates of FRT would feel safer in public places and seek social interactions they would otherwise obviate.

In an online survey (N = 6633), participants from China, Germany, the UK, and the US, were questioned about FRT and, among other things, their general acceptance of it [23]. Because of the online participation, the authors estimated them to be keener to technology and slightly younger than the average overall population. The answers regarding the general acceptance of FRT usage in the public were rather mixed: *somewhat* or *strongly accepting* were 50% of participants from China, 39% from the UK, 35% from the US, and 37% from Germany, while 22% from China, 33% from the UK, 38% from the USA and 37% from Germany were *somewhat* or *strongly opposing* it. Respondents from China showed a lot more trust in their government regarding the employment of previous surveillance techniques in comparison to western nations. Furthermore, the support of surveillance techniques implementation in comparison to opposition possessed a significantly wider gap (52% vs 13%) than in the western countries, where people were more often evenly distributed towards the ends with more people having neutral opinions. These findings correlated with their answers of how much they trust their governmental institutions in general. While a majority in all four countries expected FRT to increase the security, only participants from China saw with a larger share (>50%) improvements in efficiency and higher convenience.

Other studies were more focused on the acceptance of the implementation of FRT in consumer-orientated activities such as automatic payment systems or customer recognition benefits [31]. Such activities take place in the semi-public space, from which we can derive some new findings. In these situations, FRT is applied, for example, at the service counter where customers are required to scan their face instead of traditional authentication methods for payment authorization, such as bank cards or PINs. Commercial operators could then have access to biometric face databases to identify the person and arrange the payment transaction. The studies exposed that people indeed have concerns about privacy and security, but were overall more positive towards the adoption and usage of FRT in this context. This might at first seem surprising, but it may lead us to a key aspect of how people perceive the implementation of FRT depending on their personal choice of the technology, and the possibilities to avoid the applications. If surveillance in full public spaces is subject to civil institutions to ensure public security, people are less likely to perceive immediate benefits from it. In contrast, the automatic payment systems ease the payment process of their shopping experience and thereby deliver immediate feedback for their cost-benefit assessment. The advantages of saving time and reduced costs of items could overpower their previous discomfort with their loss of privacy. Thang and Kang [31] called this perceived usefulness, which influences whether people will finally accept the technology. Furthermore, one could argue that the slow adoption and introduction of FRT in daily activities can incrementally increase the overall public acceptance of such technologies in broader contexts like the before defined surveillance systems.

3 Facial Recognition Surveillance in Practice

3.1 Xinjiang as a Case Study

China is progressing by leaps and bounds in AI and modern technology fields. The edifice of modern China stands on cutting-edge technology put into practice. In Shenzhen's Longgang District, it took 15 h from the first police report to the location of the abducted child, thanks to facial recognition cameras [2]. South China Morning Post reports that in the same city, metro trains have cameras installed in every carriage which monitor not only all subtle moves of the passengers but also their facial expressions and transmit them in real-time as images [25].

One of the other more notable projects is Project Xueliang [24] which aims at linking up all the surveillance cameras and databases across the nation. This project's name is a play on words (meaning 'sharp'), linked to a slogan from Mao Zedong's era, 'people have sharp eyes'. It is a slogan that incited everyone to spy on everyone else, be it neighbours or loved ones.

Among other areas in China, one of the states takes the cake, Xinjiang [24]. A region with the heaviest surveillance systems implemented and a police station every few hundred meters. In this region of China, the residents are subjected to daily surveillance. To buy petrol, one must first get his face scanned at the gas station. After this scan, the system in place has to declare the subject harmless to buy fuel. Facial recognition technology that is in place records every move as the subject performs its daily tasks. If this system in place identified one as a troublemaker, an alert is sent out in most cases. On the one hand, it is better for society if all troublemakers are caught before they do something harmful, but on the other hand, for this system to function, one must assume that the one who put this system in place is all-knowing and just. For this surveillance society to be fruitful, one must believe that the algorithm that classifies people is also just. However, there is no such thing as the perfect algorithm. Therefore, the question is, is it then even remotely ethical to leave every citizen at the mercy of an algorithm that has chances of containing bias?

4 Influence on Individual Behavior

4.1 Surveillance Influences How People Act

In the Newcastle University UK, the psychology department's coffee room has an honour system in place to collect payments for the coffee: users pay for their drinks by putting their money in a box. They could also choose not to put any money in as well, because no one is watching them. Melissa Bateson and colleagues from the department put up new price lists each week for the coffee room. Prices do not change every week, but each week there is a new photocopied picture at the top of the price list. For example, one week a picture of flowers and the other week a pair of human eyes fixated directs at the paying patrons. In the weeks when eyes were displayed on the list, the people paid 2.76 times as much for their drinks as they did in comparison to the other weeks. The researchers in this study [18] were staggered by the results. According to them, a gaze of staring eyes (or a drawing of gazing eyes even) has changed people's behaviour. Attorney and former executive director of the National Lawyers guild Heidi Boghosian in her book 'I have nothing to hide' [6] talks about a myth that people have about surveillance. "Surveillance does not influence how I act". She states that the purpose of this myth is a way to normalize ongoing violations of personal privacy. The feeling of being watched proves to be a powerful tool for social control. It causes people to self-censor and conform to the norms. The coffee-room phenomenon could not have been explained if the myth "surveillance does not influence how I act" was correct. Now let us try to put this into perspective and imagine a surveillance state based on Bentham's Panopticon. To Jeremy, the subject is observing his own behaviour and disciplining himself, which at first seems like a beneficial thing for the state. A state whose citizens abide by the law because they are being watched. According to Foucault [13] one who is under the field of visibility and who is aware of it simultaneously plays two roles: the observer and the one being observed.

4.2   Hindering People to Experiment with Eccentric Ideas

The secret ballot system is the most prevalent vote casting system in the world today. The voter's identity remains anonymous as he walks into the booth to exercise his right to vote [8]. There are various advantages to voting privately rather than doing it openly, including avoiding external intimidation and the ensuing censure of the choice. As a result, the voter can cast his ballot without worrying about repercussions. Now picture a society in which everyone is required to declare their vote in public. First off, many voters today would not have cast a ballot at all out of concern for the possibility of being made to answer for their choices. Second, many individuals wouldn't vote for candidates with eccentric views or those who represent a minority even if they wanted to.

In modern liberal societies, every citizen is entitled to exercise his social liberties. This includes but is not limited to freedom of speech, the right to vote and the right to privacy. Let's now broaden this perspective and include FRT surveillance. It can become challenging for people to exercise their civil liberties, such as basic communication, social gatherings, and discussion of political topics, in a society where FRT is in existence. Because of the invasive monitoring and ongoing oversight, this would especially prevent people from experimenting with new ideas. It is simple to carry out this mental exercise and realize how many people would consider FRT to be an infringement on their civil liberties.

In the words of T. Snyder [22], *"we are free only when it is we ourselves who draw the line between when we are seen and when we are not seen"*.

In the case of Xinjiang, the extreme surveillance system targets the minority Muslim community, the Uighurs. The state forbids the practice of the religion, for example, fasting in Ramadan. The FRT is a key instrument in the Chinese crackdown against the minority [24]. China's endeavour at making perfect citizens is a form of oppression. Fear of repercussions can force citizens to act by what is considered the 'correct behaviour'. However, is there a definition of the 'correct behaviour' and especially in a diverse society, who should be entitled to specify what 'correct behaviour' is? Such a surveillance state ceases to be a peaceful and safe place for everyone. Citizens may never show their real character but the character that is in line with the 'Correct Behaviour' officially specified by the government.

4.3   Perception on Safety and Surveillance

Authorities often claim safety benefits in connection with the implementation of FRT in public spaces as discussed before. Though, the question arises of how single individuals assess this supposedly new attainment of safety, and if they evaluate it as a gain or a loss to their daily life. To answer these questions one has also to consider the previous conditions people have lived in and under which circumstances the measurements were implemented. People from areas with higher rates of violent crimes could perceive the implementation of FRT surveillance systems as again to their safety, while people from beforehand safer areas could sense it as an encroachment on their freedom with little benefits to their overall safety. Just as recent terrorist attacks and high-profile crimes can also have an influence on people's feelings of safety [23]. Overall, what people think and feel about the introductions of FRT technology with regard to their safety and what the actual benefits are backed by numerous studies in various contexts, are two different things. Independent of the controversial outcomes with mixed results as explained before, discrepancies between both perspectives could be problematic in the long run for the acceptance of such technologies and ultimately lead to their abolition. One key problem with much of the literature on surveillance in relation to the perception of safety is the sparseness in the context of FRT. Many studies investigated video surveillance without more sophisticated implementations or did not mention the exact form of data processing in the background of such systems. Because of the superior identification function of FRT and its commonly automated detection, we rate video surveillance with FRT as an expanded version of simpler systems. From there, we make the assumption, that the results of previous studies could be adopted to such new systems because participants should possess similar characteristics of attitude towards both types of surveillance levels.

Zhang et al. [30] conducted a study with a large number of participants (N = 1080) residing in Beijing with various demographic characteristics such as age, gender, education, income, occupation, and duration of residency in the city. The study aimed to comprehend the attitude of those towards the current implementation of surveillance systems regarding their own perception of safety. They found out that female respondents assessed surveillance cameras as more beneficial for their safety in comparison to male respondents, especially for crime reduction. Women stated preferences for routes and housings with such systems, saw it as less necessary to be watchful, and had an overall higher approval and acceptance for surveillance implementation. These findings align with broader studies, which compared general fear of crime gender-wise and saw significant differences in women's perception of risk, their actual risk, and their behavioural response [30]. Older people expressed more concerns about such systems such as the status-quo being sufficient safety-wise and increased stress levels as a response to the presence of surveillance systems. Surprisingly, people with more than 10 years of residency were more likely to support these measurements, perceived higher safety from it, and tend to support further expansion. This could be traceable to an effect of habituation, common for the long-term usage or, in this case, the presence of initially uncommon technology [17].

5   Influence on Collective Behaviour and Society

5.1   Privacy as a Privilege

A privilege is formally defined as *"an advantage that only one person or group of people has, usually because of their position or because they are rich"* [1]. We now want to discuss the consequences of mass surveillance of public spaces with FRT in the context of wealth inequality and the different degrees of exposure to it for people of distinct wealth affiliation. The following argument aims to picture the unequal restrictions and influences on people's lives by FRT depending on their social circumstances, especially wealth-wise. There exist numerous definitions to describe wealth and income inequality as well as methods to divide the population into such models. For our following argumentation here we would rather not be bound to specific limits of income or wealth for such allocation and will abstract from many social-cultural factors of the real world for simplicity. We will just distinguish between an upper-class elite with a high amount of social privileges [5] and a working lower- to middle-class with less amount of social privileges inspired by multi-level economic class models [28].

5.1.1   Working Lower and Middle-Class We define this stratification here as people with lower or middle-income streams and fewer possibilities to build wealth over time. They depend more on the properties of wealthier people, are in need of stable employment to finance their life expenses, and live in rental units instead of owning large properties or residential buildings themselves. Additionally, they are using public transportation more frequently and are more likely to attend public schools, depend on public hospitals, enter public institutions, and visit other forms of public or semi-public spaces. This points to the idea that they are, in general, more likely to be exposed to the surveillance of FRT and more importantly, they have less to no choice and possibilities to avoid such monitoring. They have no control over the personal data processed by such surveillance systems and have to trust the operating authorities to not abuse their power. If someone has not had the resources for owning a car or taking a cab to their workplace, they could be forced to use public transportation with the possibility of FRT surveillance. Likewise, their workplace could require specific FRT systems for measures of safety and workforce monitoring and their rental housing units require some form of FRT access agreement to enter the building. Overall,

people of this class are under the impression of constant surveillance because of their few private places which is due to the lack of their financial resources and existing dependencies on public places in their daily life.

5.1.2   Upper-Class People of this class have higher or the highest income levels, which result in accumulated wealth or receiving large inheritances, for which reason they need not pursue gainful employment. They are more likely to possess their own properties with larger estates and overall living space. In addition, they can choose their means of transportation more independently and isolated from the public space by driving their own cars. Their financial means make it possible for them to acquire more services from the private sector, which could be subject to stricter privacy rules and through this less FRT surveillance. Overall, they can collectively form some kind of privately owned area, whose access is restricted by membership payments. In certain ways, this is already implemented in the form of gated communities, most common in the US. Whereby there exists FRT technology to exclude other people from entering their residential area, inside of it could exist some form of safe space, which abandons strong surveillance systems for the local residents themselves [20]. Consequently, people of this class can withdraw themselves from FRT surveillance through building places of retreat and privacy only possible by their high amount of financial resources. Entering public places is more optional for them and they can determine the scope of possible surveillance with their own decisions regarding exposure.

What the above subdivision and consequences also try to highlight, is the increased danger of dividing the society. If FRT surveillance gains widespread adoption in public spaces, it will have a higher impact on people with no possibility of avoiding such places. People of lower-income and fewer assets have a higher dependence on public goods in general, which are connected to public spaces. FRT's introduction could similarly amplify the expansion of privately sealed-off places, where people interact, which can afford its entry. That would then again increase the separation of society and emphasize and reinforce inequality in society.

5.2   Techno-Fascism

The term techno-fascism was coined by Janis Mimura and represents a new form of authoritarian rule controlled by technocrats [19]. According to Chet Bowers, techno-fascism is characterized by the increasing reliance on digital technologies in more areas of everyday life. He emphasizes that while this reliance has many positive impacts, it also reduces the diversity of cultural knowledge and increasingly subjects human thought and behaviour to the rules of machines [7]. For instance, in Xinjiang's case, the surveillance system in place enforces the conformity of behaviour. We contend that FRT's potential to instil techno-fascist traits in the society is its most significant potential effect on society as a whole. Individualism slowly disappears and is replaced by conformity of thought and behaviour when technology like FRT surveillance is democratized throughout a society to enforce the enforcer's will. One might think that this idea is far-fetched and that with the modern democratic ideology in place, we are protected against such threats, but this is what Snyder [22] calls a 'Misguided reflex' and that one should examine the history and try to understand sources of tyranny.

5.3   Nobody Goes *Against the Grain*

As discussed before in the impacts of surveillance on individual behaviour, FRT poses a threat to society, hindering people from exercising their civil freedoms. Only such freedoms enable people to perform acts that make them stand out and advance society positively. This result particularly shapes the collective behaviour of society. Snyder stresses [22] that individuals who did exceptional deeds in the past and stood out are the ones that shaped our present drastically. Great deeds that influenced history were often carried out in opposition of existing laws. One of such cases is the case of Teresa Prekerowa [16] a young Polish girl who went to great lengths to rescue a girl she barely knew. Especially when most people would only think about themselves, she stood out by acting altruistically and helping others. She repeatedly risked herself and supplied medications and food to the Warsaw ghetto in Nazi-occupied Poland. Prekerowa ended up saving many Jewish lives, and her actions were the result of her courage and partly luck. Each time crossing, she had to encounter guards at the gate, and luckily, she was never seen by the guards outside the Warsaw Ghetto. Again, if one puts an FRT surveillance in place, it would be much harder for future rescuers like Teresa Prekerowa to gather courage. It would be infinitely more difficult to dodge the never-sleeping eye of the FRT surveillance. Thus, rescuers cannot dare to go against the grain in a society with FRT surveillance in place.

5.4   Anticipatory Obedience

Beyond the threats that have been already discussed, FRT surveillance can cause issues of psychological safety within the society which can lead to what Snyder calls 'Anticipatory Obedience' [22]. When people feel psychologically safe, they feel respected and can show their ideas, concerns, and questions without the fear of negative consequences or being humiliated or punished. However, in a society where people are self-disciplining themselves as a result of the constant watch they have to live under, they scrutinize their actions harshly. Such people may anticipate Draconian constraints on diverging behaviour, and enter a state of anticipatory obedience where they are ready to apply self-censorship and 'obey in advance'. T. Snyder elaborates that citizens start thinking ahead of time about what the government wants and how they could better align themselves to what is asked of them, without explicitly being asked. This way, the adapting citizen teaches power what it can do [22].

6   Conclusion

This paper highlights some of the implications that facial recognition technology (FRT) would have on individuals and societies as a collective. In summary, it is irrefutable that people's behaviour would alter in the presence of FRT and that some of it would, if not stop, then make it difficult for them to stand out. The individuals would self-censor and try to conform to the norms in place. FRT surveillance in place effectively realizes Bentham's Panopticon. Furthermore, if individual behaviours change, it is improbable that the collective behaviour stays unaltered. FRT has the potential to bestow certain characteristics on the society in which it is set up, that could enable the manipulation of people's behaviour. In such a society, it would be exceedingly difficult for people to perform acts that qualify as going against the grain of the 'correct' behaviour expected of the citizens, individualism would vanish and conformity is likely to take its place. Lastly, another impact of FRT can be that the society would be home to individuals who are ready to adapt and sometimes obey without even giving it any thought; thus, an environment of anticipatory obedience is likely to be created. It is, therefore, necessary to consider these consequences, before adopting a technology that holds the potential to alter the behaviour of societies.

References

1. PRIVILEGE. In: Cambridge Dictionary English. Cambridge University Press, January 2021. https://dictionary.cambridge.org/dictionary/english/privilege
2. Bandurski, D.: Big data, big concerns (2017). https://medium.com/china-media-project/big-data-big-concerns-da247a16d2f3
3. Bentham, J.: Panopticon: or, The inspection-house. Containing the idea of a new principle of construction applicable to any sort of establishment, in which persons of any description are to be kept under inspection, etc. Thomas Byrne (1791)
4. Bentham, J.: The Works of Jeremy Bentham, vol. 7. W. Tait (1843)
5. Black, L.L., Stone, D.: Expanding the definition of privilege: the concept of social privilege. J. Multicultural Couns. Dev. **33**(4), 243–255 (2005). https://doi.org/10.1002/j.2161-1912.2005.tb00020.x. https://onlinelibrary.wiley.com/doi/abs/10.1002/j.2161-1912.2005.tb00020.x
6. Boghosian, H.: I Have Nothing to Hide: and 20 Other Myths about Surveillance and Privacy. Beacon Press (2021)
7. Bowers, C.: Is the digital revolution sowing the seeds of a techno-fascist future? (2015). https://truthout.org/articles/is-the-digital-revolution-sowing-the-seeds-of-a-techno-fascist-future/
8. Brent, P.: The Australian ballot: not the secret ballot. Aust. J. Polit. Sci. (2006)
9. Bruce, V., Young, A.: Understanding face recognition. Br. J. Psychol. **77**(3), 305–327 (1986). https://doi.org/10.1111/j.2044-8295.1986.tb02199.x. https://bpspsychub.onlinelibrary.wiley.com/doi/abs/10.1111/j.2044-8295.1986.tb02199.x
10. Brunelli, R., Poggio, T.: Face recognition: features versus templates. IEEE Trans. Pattern Anal. Mach. Intell. **15**(10), 1042–1052 (1993). https://doi.org/10.1109/34.254061
11. Chellappa, R., Wilson, C., Sirohey, S.: Human and machine recognition of faces: a survey. Proc. IEEE **83**(5), 705–741 (1995). https://doi.org/10.1109/5.381842
12. Consejo, D.E.: European convention on human rights. In: European Convention on Human Rights. https://www.echr.coe.int/Documents/Convention_ENG.pdf
13. Foucault, M.: Discipline and Punish. Pantheon, New York (1977). Sheridan, A. (trans.)

14. European Court of Human Rights: Guide on article 8 of the European convention on human rights. Right to respect for private and family life, home and correspondence (2019)
15. Introna, L., Nissenbaum, H.: Facial recognition technology: a survey of policy and implementation issues (2009)
16. Kalabinski, M., et al.: Teresa Prekerowa: going against the grain. East. Eur. Polit. Soc. **34**(02), 400–422 (2020)
17. Liao, C., Palvia, P., Chen, J.L.: Information technology adoption behavior life cycle: toward a technology continuance theory (TCT). Int. J. Inf. Manage. **29**(4), 309–320 (2009). https://doi.org/10.1016/j.ijinfomgt.2009.03.004. https://www.sciencedirect.com/science/article/pii/S0268401209000292
18. MacKenzie, D.: Big brother' eyes make us act more honestly. New Sci. **28** (2006)
19. Mimura, J.: Planning for empire. In: Planning for Empire. Cornell University Press (2011)
20. Norazizi, N.S.: Security systems in Residential Areas. IRC (2020)
21. The Internet Encyclopedia of Philosophy: Surveillance ethics (1999). https://iep.utm.edu/surv-eth/#:~:text=Surveillance%20is%20itself%20an%20ethically,employed%2C%20and%20questions%20of%20proportionality
22. Snyder, T.: On Tyranny: Twenty Lessons from the Twentieth Century. Random House (2017)
23. Steinacker, L., Meckel, M., Kostka, G., Borth, D.: Facial recognition: a cross-national survey on public acceptance, privacy, and discrimination (2020)
24. Strittmatter, K.: We Have Been Harmonized: Life in China's Surveillance State. HarperCollins (2020)
25. Tao, L.: Metro pickpockets beware! China's hi-tech cameras are watching your every move (2018). https://www.scmp.com/tech/enterprises/article/2126553/metro-pickpockets-beware-chinas-hi-tech-cameras-are-watching-your
26. Turk, M., Pentland, A.: Eigenfaces for recognition. J. Cogn. Neurosci. **3**(1), 71–86 (1991). https://doi.org/10.1162/jocn.1991.3.1.71
27. Wheeler, F., Liu, X., Tu, P.: Handbook of Face Recognition, 2nd edn. (2011)
28. Wright, E.O.: Class Structure and Income Determination, vol. 2. Academic Press, New York (1979)
29. Yang, M.H., Kriegman, D., Ahuja, N.: Detecting faces in images: a survey. IEEE Trans. Pattern Anal. Mach. Intell. **24**(1), 34–58 (2002). https://doi.org/10.1109/34.982883
30. Zhang, H., Guo, J., Deng, C., Fan, Y., Gu, F.: Can video surveillance systems promote the perception of safety? Evidence from surveys on residents in Beijing, China. Sustainability **11**(6) (2019). https://doi.org/10.3390/su11061595. https://www.mdpi.com/2071-1050/11/6/1595
31. Zhang, W., Kang, M.: Factors affecting the use of facial-recognition payment: an example of Chinese consumers. IEEE Access, 1 (2019). https://doi.org/10.1109/ACCESS.2019.2927705
32. Zhao, W., Chellappa, R., Phillips, P.J., Rosenfeld, A.: Face recognition: a literature survey. ACM Comput. Surv. **35**(4), 399–458 (2003). https://doi.org/10.1145/954339.954342

Enhancing Transparency of Political Micro-targeting on Facebook

Florian Medert[1], Jan Fridtjof Otto[2], and Léna Perczel[3(✉)]

[1] European Research Center for Information Systems, University of Münster, Münster, Germany
florian.medert@uni-muenster.de
[2] Science Policy Research Unit, University of Sussex, Brighton, UK
j.otto@sussex.ac.uk
[3] Political Freedoms Program, Hungarian Civil Liberties Union (HCLU, TASZ), Budapest, Hungary
perczel.lena@tasz.hu

1 Political Micro-targeting and the Problems It Raises

The importance of political micro-targeting is undeniable. One of the main points of democracy is to have a political discourse that helps in the advancement of ideas and the exchange of values. An important part of enhancing the protection of political discourse is the protection of the freedom of speech. Nowadays, the political discourse is highly affected by social media, as being the platform where people communicate and exchange their thoughts and values. Social media, modern search engines and targeted advertisements profoundly influence one's perspective on such questions [59]. One particular form of this influence is online political micro-targeting. Online political micro-targeting "involves' creating finely honed messages targeted at narrow categories of voters' based on data analysis' garnered from individuals' demographic characteristics and consumer lifestyle habits" [84]. Thus seeing the threat of micro-targeting raises the question of regulating it.

"Political micro-targeting relies on the sophisticated psychological and technological methods, developed by the commercial advertising industry, of collecting information about users' preferences and organising them into user profiles to target them with personalised messages" [7]. The use of political micro-targeting is multi-purpose. It can be directly or indirectly related to political processes, lik e persuading voters, encouraging and discouraging election participation. These types of ads influence people more because these adapt to the needs and interests of the targeted users [84]. Both the collection of high amounts of personal information and the way of influencing people through targeted advertisements raise questions concerning individual rights [7].

Although micro-targeting raises many concerns, it also holds several promises, such as strengthening democracy through increasing political participation. Furthermore, it could also increase the political knowledge of citizens, thus helping them make better choices by being well-informed [84]. Therefore,

J. Bayer and C. Grimme (Eds.): AI, Human Rights and Ethics, LNAI 14400, pp. 47–62, 2025.
https://doi.org/10.1007/978-3-031-52082-2_4

making political micro-targeting more ethical and legal not only protects individuals' rights, but also enhances their capability to exercise their rights. Our aim is to discuss the requirement of transparency - as a response to new online social harms [54] - in the context of the political micro-targeting of Facebook, and to give a proposal on how to achieve this goal.

The requirement of transparency means that on the one hand, users must receive information on how their data is used and why they are being targeted, on the other hand, the information collected for political ad campaigns must be made available to public interest watchdogs: such as independent journalists and researchers [19].

There are multiple reasons why we focus on Facebook in our research. Firstly, it has an enormous number of users, and thus it has become "a key environment for political campaigns" [65]. The fact that Facebook advertisements are hard to distinguish from general posts even strengthens this notion [43]. Secondly, because Twitter and Spotify have banned, Google has restricted political advertisements on their platform as of 2019 [65]. And although Facebook has limited some of its ad-targetings, it only started the restrictions in February 2022, and the decision came after emphasising how they would not want to be in a position of deciding whether an ad should and how it should reach the voters. "The move isn't a total ban on political ad-targeting, in that it still allows campaigns to use publicly available data or their own email lists as ways to reach people on Facebook. Political strategists said as a result the social-media giant would remain an essential tool for organizing" [77].

As an attempt to reach transparency, Google, Twitter and Facebook created ad libraries with which they could analyse online political campaigns better [53]. Ad libraries are "publicly accessible databases with an overview of political advertisements featured on their services" [47]. These public archives document the advertisements and the associated data of the users, and further information like expenditure and a number of views. Although advertisements encompass both commercial and political advertisements, in practice, the ad libraries focus on political advertisements [47].

According to Mehta and Erickson, the goals wished to achieve through establishing the Facebook Ad Library (FAL) with regard to political advertisements are undermined by its limitations. Leerson et al. argue three main points why ad libraries proved to be insufficient in regard of reaching transparency. Firstly, ad archives faced difficulty in defining "political advertisements" thus identifying them, as there is no commonly accepted definition. Secondly, ad archives have proven vulnerable to inauthentic behaviour, particularly from ad buyers seeking to hide their true identity or the origin of their funding. Thirdly, they do not hold sufficient and vital data on how ads are targeted and distributed [47].

We can see how multiple scholars have pointed out that the existing ad libraries do not fulfil their aim. Therefore, self-regulation applied by the giant platforms becomes questionable as an effective way to step up against the emerging issues in the online environment. Ad libraries have been a promise for regulators and policymakers, but apparently the requirements articulated proved to be insufficient. We will look at what the most relevant parts of the regulatory framework look like now, and how they should and could be improved.

Since 2018, many governments all around the world started to address the issue around political micro-targeting and the responses of the platforms such as Twitter, Google, and Facebook were diverse. In some regions, they have sought to comply with the obligations imposed on them, but in some others they have refused to and decided to ban political advertisements in those regions completely [47].

In the last few years, the European Union started to take the online risks posed by social media platforms very seriously. It drafted several legal instruments and policy guidelines that addressed these risks, such as the 'Ethics Guidelines on Artificial Intelligence', the 'AI Act', the 'Digital Services Act' (DSA) and the widely applied 'General Data Protection Regulation' (GDPR) which became an example for the regulation of Data Protection. The current most specific one regarding political micro-targeting is the 'Proposal for a Regulation of the European Parliament and of the Council on the transparency and targeting of political advertising' published by the European Commission on 25 November 2021.

According to the proposal, there is a fragmentation of transparency requirements due to the different definitions of political advertising and the inability to keep up with the rapid technological changes. Thus, it states that "the rapid technological change, increasingly fragmented and problematic regulatory context, and the increasing amounts being spent on political advertising demonstrate the need to act at the EU level to ensure the free movement of political advertising services across the Union while ensuring the high standard of transparency which makes electoral processes in the EU more open and fair" [17].

The proposal received criticism due to the lack of real-time transparency, which would be "essential in a fast-paced electoral environment". Furthermore, it does not provide meaningful transparency on advertising campaigns as a whole, only on individual ads. Thus, the accountability of campaigns by watchdogs and media is unreachable. And not individual ads, but multiple can cause the most violations of electoral rules [28].

Although the proposal's aim is to unify regulation surrounding political advertisements, partly by providing a uniform definition, according to Lindroos-Hovinheimo the definition provided is still too broad [49]. Thus, some are concerned that it could significantly impact freedom of expression. The ambiguity of the definition does not differentiate clearly between editorial content and political advertising, and thus it could limit political reporting as well [18], which should not be the aim of the proposal. We can see that the existing regulations and standards that address the issue lack specificity, which hinders their effectiveness.

Since people need to know why and how their data is used (Chang, 2020 [15]), public policy experts and major regulators must find better solutions to the regulation of political micro-targeting to meet the needs of the citizens. We will attempt to do so too, but first we unfold the underlying technological system, and then assess micro-targeting's effects on users. Afterwards, we provide an insight into the existing proposals in literature and then finally give our own proposal.

2   The Underlying Technological System: Machine Learning for User-Profiling

The following chapter gives a general overview of what machine learning and artificial intelligence is and how it can be used to analyse "big data" sets and how the analysis of user data can be used to create profiles for targeted advertisements.

In this chapter, "the advertiser" or "the advertising firm" is used to describe the company or group that deploys the advertisement to the user. In most cases, specifically in social media ads, this is not the institution provisioning advertisements for its product or political campaign, but rather the social media platform that sells advertisements.

By definition, machine learning is "the study of computer algorithms that improve automatically through experience" [58, p. 1]. In this study, the algorithms are improving their understanding of user data through experience (labelled user-data and newly generated user data). For the context of this study, the concept of machine learning as the underlying technology is presented in general terms to show how machines (can) create user profiles based on sets of user data, to analyse user interaction, and to, in turn, use these profiles to tailor targeted advertisements. That is, representing (user) data with multiple levels of abstraction [46], e.g., using user profiling to identify target segments of audiences for (political) advertisements.

A company aiming to deploy a personalised and targeted (political) advertisement to a user initially needs two things: first, it needs to define its target, as in the user or audience segment it wants to deploy the advertisement to. Second, for the ad to be personalised/targeted, it needs an ad that fits the target user in a predefined way. For example, if the target user is specifically interested in a specific sport, e.g., ski, the target advertisement may be for ski, or ski attire. In this chapter, both aspects are of interest, but in practice, the former must be fulfilled first. As it is unfeasible to select each individual user to deploy advertisements to, the advertiser may want to choose a "type of user" based on a set of characteristics, thus, a profile.

First, according to van Otterlo [64], it is necessary to distinguish between persons (users) and profiles. The person is the user, the human interacting with/through the social media platform. A profile is used to describe "models, that model correlations between pieces of information appearing in the individual's data, causal patterns and general rules that apply to a subset of the individual" [64, p. 4]. The profile, thus, infers knowledge about the person that is not necessarily observable. The person "fits" into multiple profiles (or at least some aspects of them) [64, p. 4]. Thus, the advertiser deploys the ad to users who fit the profile and their target audience, rather than finding all users individually. This concept can be applied to more than one profile, as users fit multiple profiles simultaneously. More conclusively "a user profile is a set of information representing a user via user related rules, settings, needs, interests, behaviors and preferences" [20].

User profiles consist of/are built from user content. This content may be static (e.g., demographic information like 'age') or dynamic (e.g. behaviour, like pressing the 'like' button on a post) [20,31]. The information (data) used in user profile construction may be gathered using explicit or implicit approaches. A combination is possible too [31]. The user provides explicit information directly and willingly, e.g., by entering his age in the registration form. Implicit information is gathered through the user's interaction with the platform (e.g., liking a post) [20]. Thus, whenever a user interacts with the social media platform in some way, a datapoint can be created. This includes search logs, access logs, browsing history and even the usage of links to external sites [31]. Generally, it can be assumed, that profiles on social media platforms are created using a mixture of explicit and implicit data gathering approaches, thus containing information given both intentionally, as well as unintentionally. Facebook, for example, publicly states, that "as everyday millions of users use [their] apps and interact with content, the system gets a lot of information to improve its calculations" [30].

The gathered data can then be used to construct user profiles using different approaches. While information retrieval approaches (using term and topic vectors to link users to interests), ontology-based approaches (using reference ontologies to link user interests to topics) or agent-based approaches are used in the past, as well as in recent studies [31]. This study focuses on machine learning approaches. The approach "uses machine learning techniques to model user behaviour" [31, p. 5]. Methods may be decision trees, Naïve Bayesian classifiers and K Nearest Neighbor algorithms based on a labelled set of training data [31]. Training data for this may include explicitly gathered user information, like surveys. For example, in the Cambridge Analytica case, a psychological survey conducted by researcher Aleksandr Kogan was used as the base training data to construct profiles for political partisanship that were later used to link millions of US citizens too [52,66]. This approach is specifically used to analyse implicit user data, that is potentially of lower quality but higher quantity. User interactions with the platform, e.g., signal interests, are dynamic data with high frequency, thus requiring more automated methods [20, p. 2].

In the wake of the Cambridge Analytica case and the subsequent discussion about political advertisement and targeted advertisements, an additional distinction presents itself. The difference between demographics and psychographics and the resulting personality profile emerges. While demographics and geographics are frequently used to segment audiences into e.g., age, gender or nationality groups, psychographics or psychometrics convey a deeper understanding of the user's personality and, thus, a clearer insight into intricate personal behaviour, e.g. political partisanship [61]. Nix states, that "demographics are important, but equally or even more important is psychographics: an understanding of your personality, because it is personality that drives behavior, and it is behavior that influences how you vote" [61].

The psychographic perspective, namely the perspective of the "Big Five Model" frequently called "the OCEAN" model finds traits, that all humans share beyond demographics [62]. These traits are openness (how open is a user to new experiences), conscientiousness (competence, order, self-discipline), extraversion (how social is the user), agreeableness (respect to the needs of society/community) and neuroticism (tendencies to worrying) [61,62].

Based on a training dataset (e.g., the surveys by Aleksandr Kogan), machine learning algorithms may use data mined from social media platforms to construct user profiles and continuously fit any individual that interacts with social media platforms [66, p. 76]. Thus, the users are targetable as intended for targeted advertisements.

3   Effects on Users (Humans)

The existence of targeted, especially micro-targeted advertisement based on profiling has many effects on humans and many (if aware) implications for the recipient of such ads.

Initially, this lowers the trust in democracy [15], as political ads feel constructed to persuade the user, rather than state a position the user may agree or disagree with. It induces unclear perceptions of what the party actually aims for, and what their goals are. Also, this makes it unclear if the resulting mandate has the voters' approval. Because targeted ads are hidden from everyone but the target user/segment audience, there is little possibility to challenge the information (Haewood et al., 2018), and hold the advertisement accountable for e.g., misinformation [47]. Ads may (are) thus able to include lies or at least misleading information [63]. It also may differ depending on culture, identity, gender and race [3,4,57]. Political micro-targeting is shown to have worked in the past [1], and is especially used in less clearly polarised areas and districts [33,34]. For example, in US elections, it is primarily the "swing-states" (states not clearly republican/democratic-sided) that sway the final overall election results. Methods can both be positive or negative toward a candidate or party (Kaid & Boydston 2009 [40], Shen Wu 2002 [69]). Finally, user groups who are not targeted are excluded from public discourse, which opens the question of a fundamental right to receive information and to be informed about political positions and goals [7,22].

Overall these effects have major implications for a democracy. If political discourse is fragmented to specifically target groups based on profiling, a democracy sways from the idea of collectively deciding on issues by voting for a party and their goals to voting based on what a party or group wants you to think their goals are wrapped in a perfectly tailored persuasion package. As social media is one of the key aspects in our lives where humans spend a significant amount of time and receive potentially a major part of their advertisement, the emerging form of political targeted ads is unsafe for democracy and conveys major ethical questions. In the following we identify current proposals for designing this concept and show our own proposal on the topic.

4   Current Proposals and Ideas in Literature

Most proposals that can be found in literature see the obligation to propose effective solutions in the hands of politics. This mostly can be traced back to Facebook's current poor rather opaque attempt at self-regulation and its history of various repeated scandals when it comes to targeted advertisements (ads) [2, 9,35,53,73,78]. Some proposals in the literature target only a little aspect of the described problem such as Venkatadri et al. [79] or Liu et al. [50]. These and similar unscalable approaches that focus on unscalable sub-problems have been excluded from this paper for conciseness.

The first joint proposal in literature is to generally pick up Facebook's attempt of assembling an Ad library and to extend the provided content extensively. It has been noticed that the Facebook Ad Library (FAL) is not effective in providing researchers and journalists and electoral monitoring authorities with the ability to monitor a campaign [25]. Precisely, Mehta and Erickson [54], similar to what is proposed in articles 30 and 31 of the digital services act, demand disclosure of the author of the ad and their affiliation to an organization/political standpoint, as well as indicating clearly whether micro-targeting was used to place the ad [68]. To facilitate this, a key requirement is to collect this data adequately. Currently, it is still frequently the case that political ads are not identified as such [45]. One reason here is that advertisers need to self-declare their political ads [72]. Further Facebook's unclear definition of "political advertisements" results in a big share of political ads not being shared on the FAL or not recognized by the algorithm [73]. This is further complicated by the currently proposed definitions in literature having distinct pitfalls, calling for clear legislation [37]. Some suggest a practice of an external party auditing the correctness of the FAL to ensure that it reflects reality correctly [72]. The same lack of adequateness applies to who is placing the ad. To place a political ad only a photo of your ID and the name of your company is needed. In the past, algorithms used to verify ads, still allowed users to place false names in the "paid for by" section of the advertisements. Various scholars demand such gaps to be filled [10,21,82]. Shiner [71] proposes to separate political advertisements from the normal feed to increase the clarity that a specific piece of content is a political advertisement to increase awareness when seeing an advertisement. Additionally, it is also suggested to increase the clarity of what is an advertisement in general, since sponsored political content and its origin is frequently not registered as such by users [11].

Others suggest unifying all existing ad libraries of different social media platforms into one overall ad library that is set up on a European level, facilitating the possibility to analyse political ads across platforms [6,26,29,38]. A key requirement demanded by literature for such an ad library would be to continue facilitating API access for mass data analysis [60]. Another main point of critique with the FAL is that the data is not provided on the platform in real-time, but is partly uploaded only at a later point in time. Real-time transparency is crucial in the fast-moving world [42]. Therefore, improving the Facebook's timeliness to add Data to the FAL is a frequent demand [16]. Additionally, some suggest adding an analysis tool directly to the FAL, which can provide users, scientists or reporters directly with overarching trends, commonalities and connections, similar to the political dashboard proposed by Serrano et al. [70]. The calls to put an independent third party or governmental party in charge of such an ad library were further enhanced by some ads on the FAL being removed in retrospect and Facebook not providing information of for instance, ads that were undisclosed by the users and therefore got removed [27].

Concerning AI, transparency can be implemented on various levels (Input Data, Algorithm, Goals, Outcomes). In our context, transparency implies that the users should be informed about the methods and data that are used to arrive at a targeted advertisement [59]. Therefore a simple code review or an explanation of how the profiling algorithm works is not sufficient, but rather the use of the resulting profile data is essential. Our case goes beyond the auditability or explainability of the algorithm [75], but focuses more on how the resulting profile is used. The user should be able to see from which data points the algorithm derived a specific categorization of them and who selected this categorization to target them with a specific advertisement. Dommett [24] describes this as Source-, Data- and Targeting – transparency. The fourth aspect that Dommett identified is termed Financial transparency. Beyersdorf [10], Bohman [12] and Wood [83] describe how the public wants clear insights into the expenditures of political campaigns in online advertisements, postulating that such information should be provided clear and concisely in the FAL. Others even suggest putting a legal limit on expenditures for a campaign or restricting political ads close to an election [84]. The effectiveness of the former proposal suffers due to not including third parties, which could continue placing ads online and avoid the ban, while the latter is frequently opposed by citizens due to limiting the right of expression.

5   Our Two-Folded Proposal

Based on all the described proposals, considerations and ideas, we created an elaborate proposal covering both an ad library as well as the provision of advertisements.

We envision a European repository that should contain all ads that were placed on any social media platform on the internet to avoid the previously described complications of defining "political advertisements" and the non-disclosure from customers [37]. Considering the newly proposed definition, specifically, the latter argument is fundamental [17]. By providing all data, it can be more easily verified whether all political advertisements have been identified and disclosed as such. The responsibility for providing correct data should be clearly assigned to the platform provider, while it should be the responsibility of the EU commission to facilitate a repository. This assigns a more apparent responsibility instead of continuing the current multi-stakeholder liability [8]. All platform providers should be required to provide the following data to the EU commission: Who submitted the ad? With which organization are they affiliated? How much is spent on the campaign? Were micro-targeting practices used? (For instance, were there alternative or similar advertisements, those should be cross-referenced.) Which filters were selected to target the user (or was an external dataset used)? This information may require large amounts of data storage, as the necessary data accumulates over time. In a broader study a fitting infrastructure model for this data may be explored. For example, the recently proposed GAIA-X (cloud-)infrastructure could be used to implement this storage. GAIA-X employs transparent GDPR conform data storage and is managed by an EU initiative [32].

To ensure that the submitting party is legitimate, we suggest that Facebook improves their verification process before a user can place advertisements. Specifically, an in-person verification of their identity similar to IDnow [36] could be feasible and effective. Most of the previously mentioned data can be automatically gathered from the data provided to create the ad. This should be verified by an external monitoring system, monitoring both the correctness of the data provided by the advertiser and the correctness of the data provided by the platform provider itself. Such a system can partly function through citizen science by verifying data manually from the ad library, partly consist of inspections by an auditing committee and be further supported by algorithmic practices. Such auditing committees, hosted by the European Commission, and their equipment could be cross-financed by requiring the platform providers to provide the needed resources. The described measures would now enable NGOs to verify that the advertisement platforms report all ads to the central repository and their correctness [38,50]. The repository should fit the needs of both monitoring institutions, scientists, journalists, and platform users. This encompasses API access for data extraction, an easy-to-use but extensive search tool as well as filters. Additionally, it should provide postprocessing of the data, that identifies further relevant information from the advertisement that is not natively provided. For instance, a range of algorithms should be implemented that for instance determines the content of the advertisement, conducts a sentiment analysis of the content or identify the ideology and affiliation of the sponsor [44,67,80]. Further, this repository should be frequently improved if users suggest effective additions. If this additionally derived data from the algorithms is effectively processed and displayed in the search engine, we provide more information that is easily localisable and the ability to draw more accurate conclusions. Providing this information, termed by Michener and Bersch [55] as visibility and verifiability, increases the transparency for all people that are interested in the repository and not only skilled scientists.

To further increase the transparency for the average layperson whilst using the social media platform, we suggest also requiring a new User Interface (UI) when showing an advertisement to a user since the current UI is not effective in revealing the identity of the political advertisement and its origin [11]. Platform providers generally aim to design their interface in a way that provides as little information as possible, if the ongoing process is perceived negatively by users [76]. Specifically in regards to sensitive topics such as privacy or advertisements, a clearer UI that gives more information about the ongoing processes, leading to more attention to the negatively perceived practice, can lead to users stopping the use of the platform. For instance, by making the "Sponsored" text small, Facebook ensures that fewer users identify the content as non-organic. Such practices leading to specific behaviours of the users are termed Dark patterns and should be avoided to create transparency [56]. Specifically, we recommend increasing the size and changing the colour of the "Sponsored" text as well as the text highlighting who paid for the advertisement. Additionally, the "Why am I seeing this ad?" - button should be directly visible on the advertisement without clicking on the "..." button. This gives additional, but not too overloading information to the users, creating a level of transparency that is generally received positively by users [23]. When clicking this button, a simplified explanation of the profiling process and result should be given: "The party X has selected you as the audience of this advertisement, since you are classified as having positive sentiments towards the EU. Our profiling algorithm derived this by utilizing data points such as you frequently interact positively with Pro EU stakeholders". In this matter, by utilizing an example, better insight is given into which data the profiling algorithm used and how the outcome was used, while not disclosing the setup of the algorithm, which is a crucial trade secret. This should be done visually, for instance by using flowcharts to increase clarity [39]. Such an approach of giving a personalised explanation that gives insight into the main factors involved in the profiling process is a practical example of a 'Local explanation' which is both feasible and in line with the GDPR requirement of explainability of algorithms [13]. From this page, accessing the central repository to find more details on the specific advertisement should be made easy as well. Additionally, it should provide a summary of recent ads that targeted the specific user similar to "Who targets me" (Who Targets Me, 2022). In this way, we can enhance transparency through explainability [74,75]. To conclude, this approach enhances transparency as a facilitator of the lacking accountability, creating a system of digital checks and balances on a wider scale [48,83]. Further, it also allows the individual layperson to recognize, understand and uncover who is intentionally targeting the respective user, creating a more reflective and positive experience [5].

We believe that iterating the advertisement placement, despite potentially decreasing the effectiveness of targeted advertisements, is crucial to increase the previously described unreflected interactions of users with political advertisements. We see the need to go beyond increasing the awareness of only voluntary and interested users similar to the idea of Lorenz-Spreen et al. [51].

The exact effect of such a proposal remains to be determined and does not necessarily have to be bad, since transparency can also positively affect trust and ad effectiveness [41]. With this proposal, we tried to arrive at a balanced idea that would not intercept the freedom of expression by for instance limiting or outright forbidding targeted advertisement but still mitigate the previously described problems with targeted advertisements, which would most likely be legally feasible [81]. Nevertheless, our proposal would still be very hard to push into realisation, since the Digital Services Act, which demanded way less than we suggest, has been aggressively undermined by high lobby pressure [14]. Another limitation is that our proposal is not able to fully solve the problem of the right of non-targeted users to receive information [7]. We are looking forward to seeing whether and how this proposal will be picked up and further ideated upon.

References

1. Zarouali, B., Dobber, T., De Pauw, G., de Vreese, C.: Using a personality-profiling algorithm to investigate political microtargeting: assessing the persuasion effects of personality-tailored ads on social media, August 2022. https://doi.org/10.1177/0093650220961965

2. Angwin, J., Varner, M., Tobin, A.: Facebook enabled advertisers to reach 'Jew haters'. https://www.propublica.org/article/facebook-enabled-advertisers-to-reach-jew-haters?token=_VFbwXcbLCB2ey6jaMht7T02FN5H4bYF. proPublica

3. Appiah, O.: Ethnic identification on adolescents' evaluations of advertisements. J. Advertising Res. **41**(5), 7–22 (2001). https://doi.org/10.2501/JAR-41-5-7-22. https://www.journalofadvertisingresearch.com/content/41/5/7

4. Appiah, O., Liu, Y.I.: Reaching the model minority: ethnic differences in responding to culturally embedded targeted- and non-targeted advertisements. J. Current Issues Res. Advertising **31**(1), 27–41 (2009). https://doi.org/10.1080/10641734.2009.10505255

5. Barbosa, N., Wang, G., Ur, B., Wang, Y.: Who am I? A design probe exploring real-time transparency about online and offline user profiling underlying targeted ads. Proc. ACM Interact. Mob. Wearable Ubiquit. Technol. **5**(3), 1–32 (2021). https://doi.org/10.1145/3478122

6. Barocas, S.: The price of precision: voter microtargeting and its potential harms to the democratic process. In: Proceedings of the First Edition Workshop on Politics, Elections and Data, pp. 31–36 (2012). https://doi.org/10.1145/2389661.2389671

7. Bayer, J.: Double harm to voters: data-driven micro-targeting and democratic public discourse. Internet Policy Rev. **9**(1) (2020). https://policyreview.info/articles/analysis/double-harm-voters-data-driven-micro-targeting-and-democratic-public-discourse

8. Bayer, J.: New EU rules on political advertising: here you go, read the fine print - Judit Bayer. Inforrm's Blog (2021). https://inforrm.org/2021/11/30/new-eu-rules-on-political-advertising-here-you-go-read-the-fine-print-judit-bayer/

9. B.B.C.: Fake Cambridge analytica ad hits Facebook. BBC News (2018). https://www.bbc.com/news/technology-46043578

10. Beyersdorf, B.: Regulating the most accessible marketplace of ideas in history: disclosure requirements in online political advertisements after the: election. Calif. L. Rev **107**, 1061 (2016)

11. Binford, M., Wojdynski, B., Lee, Y.I., Sun, S., Briscoe, A.: Invisible transparency: visual attention to disclosures and source recognition in Facebook political advertising. J. Inf. Technol. Politics **18**(1), 70–83 (2021). https://doi.org/10.1080/19331681.2020.1805388
12. Bohman, C.: Cyber security - how can the Swedish government prevent political micro-targeting from threatening the electoral process? A policy memo security, policy and strategy in cyber space. https://www.academia.edu/37352763/CYBER_SECURITY_How_can_the_Swedish_Government_prevent_Political_Micro_Targeting_from_threatening_the_Electoral_Process_A_Policy_Memo_Security_Policy_and_Strategy_in_Cyber_Space
13. Brkan, M., Bonnet, G.: Legal and technical feasibility of the GDPR's quest for explanation of algorithmic decisions: of black boxes, white boxes and fata morganas. Eur. J. Risk Regul. **11**(1), 18–50 (2020)
14. (CEO): How the European Parliament's proposals on surveillance advertising changed over time (2022). https://datawrapper.dwcdn.net/2bYkD/1/
15. Chang, Y.H.: Personalized political Facebook advertisements (2020). http://othes.univie.ac.at/63249/
16. Colliver, C.: Cracking the code: an evaluation of the EU code of practice on disinformation. Institute for Strategic Dialogue (2020). https://www.isdglobal.org/Wp-Content/Uploads/2020/06/Isd_Cracking-the-Code.pdf
17. European Commission: Proposal for a regulation of the European parliament and of the council on the transparency and targeting of political advertising (2021). https://ec.europa.eu/info/law/better-regulation/have-your-say/initiatives/12826-Political-advertising-improving-transparency_en
18. European Publishers Council: Position of the European publishers council on the regulation on transparency of political advertising com (2021) 731 final. https://www.epceurope.eu/post/epc-position-of-the-on-the-regulation-on-transparency-of-political-advertising
19. Crain, M., Nadler, A.: Political manipulation and internet advertising infrastructure. J. Inf. Policy **9**, 370–410 (2019). https://doi.org/10.5325/jinfopoli.9.2019.0370
20. Cufoglu, A.: User profiling-A. Short Rev. **108**(3), 1–9 (2014)
21. Dobber, T., Kruikemeier, S., Goodman, E., Helberger, N., Minihold, S.: Effectiveness of online political ad disclosure labels: empirical findings. Institute for Information Law, University of Amsterdam (2021)
22. Dobber, T., Vreese, C.: Beyond manifestos: exploring how political campaigns use online advertisements to communicate policy information and pledges. Big Data Soc. **9**, 205395172210954 (2022). https://doi.org/10.1177/20539517221095433
23. Dogruel, L.: Too much information !? Examining the impact of different levels of transparency on consumers' evaluations of targeted advertising. Commun. Res. Rep. **36**(5), 383–392 (2019). https://doi.org/10.1080/08824096.2019.1684253
24. Dommett, K.: Regulating digital campaigning: the need for precision in calls for transparency. Policy Internet **12**(4), 432–449 (2020). https://doi.org/10.1002/poi3.234
25. Dubois, P., Arteau-Leclerc, C., Giasson, T.: Micro-Targeting, Social Media and Third Party Advertising: Why the Facebook Ad Library Cannot Prevent Threats to Canadian Democracy (SSRN Scholarly Paper (3817971)) (2021). https://papers.ssrn.com/abstract=3817971
26. Edelson, L., Chuang, J., Franklin Fowler, E., Franz, M., Ridout, T.: Universal digital ad transparency (SSRN Scholarly Paper (3898214)) (2021). https://doi.org/10.2139/ssrn.3898214

27. Edelson, L., Lauinger, T., McCoy, D.: A security analysis of the Facebook ad library (2020)
28. European Partnership for Democracy: EPD reaction to the commission proposal for a regulation on the transparency and targeting of political advertising. European Partnership Democracy (EPD). https://epd.eu/wp-content/uploads/2022/03/reaction-to-the-commission-proposal-for-a-regulation-on-the-transparency-and-targeting-of-political-advertising_25_11_2021.pdf
29. E.P.D.: Virtual insanity: the need for transparency in digital political advertising. European Partnership for Democracy (EPD). https://epd.eu/virtual-insanity/
30. Facebook: Wie setzt Facebook maschinelles Lernen bei der Anzeigenauslieferung ein?. https://de-de.facebook.com/business/news/good-questions-real-answers-how-does-facebook-use-machine-learning-to-deliver-ads/
31. Farid, M., Elgohary, R., Moawad, I., Roushdy, M.: User profiling approaches, modeling, and personalization. SSRN Electron. J. (2019). https://doi.org/10.2139/ssrn.3389811
32. G.A.I.A.-X.: Gaia-x: policy rules and architecture of standards. https://www.bmwk.de/Redaktion/EN/Publikationen/gaia-x-policy-rules-and-architecture-of-standards.html
33. Haenschen, K.: The conditional effects of microtargeted Facebook advertisements on voter turnout, August 2022. https://doi.org/10.1007/s11109-022-09781-7
34. Haenschen, K., Jennings, J.: Mobilizing millennial voters with targeted internet advertisements: a field experiment. Polit. Commun. **36**(3), 357–375 (2019). https://doi.org/10.1080/10584609.2018.1548530
35. Hern, A.: Facebook exempts political ads from ban on making false claims. The Guardian (2019). https://www.theguardian.com/technology/2019/oct/04/facebook-exempts-political-ads-ban-making-false-claims
36. IDnow: identity shop-identity verification in a store nearby (2022). https://www.identity.tm/gb/products/shop.html
37. Jaursch, J.: Defining online political advertising. Stiftung neue verantwortung (2020). https://www.stiftung-nv.de/en/publication/defining-online-political-advertising
38. Jeong, U., Ding, K., Liu, H.: FBAdTracker: an interactive data collection and analysis tool for Facebook. https://doi.org/10.48550/arXiv.2106.00142. Advertisements (arXiv:2106.00142). arXiv
39. Jin, Y., Seipp, K., Duval, E., Verbert, K.: Go with the flow: effects of transparency and user control on targeted advertising using flow charts. In: Proceedings of the International Working Conference on Advanced Visual Interfaces, pp. 68–75 (2016). https://doi.org/10.1145/2909132.2909269
40. Kaid, L.L., Boydston, J.: An experimental study of the effectiveness of negative political advertisements. Commun. Q. **35**(2), 193–201 (1987). https://doi.org/10.1080/01463378709369680
41. Kim, T., Barasz, K., John, L.: Why am I seeing this ad? The effect of ad transparency on ad effectiveness. J. Consum. Res. **45**(5), 906–932 (2019). https://doi.org/10.1093/jcr/ucy039
42. Koops, B.J.: On decision transparency, or how to enhance data protection after the computational turn. In: Privacy, Due Process and the Computational Turn. Routledge (2013)
43. Kruikemeier, S., Sezgin, M., Boerman, S.: Political microtargeting: relationship between personalized advertising on Facebook and voters' responses. Cyberpsychol. Behav. Soc. Network. **19**(6), 367–372 (2016). https://doi.org/10.1089/cyber.2015.0652

44. Kulkarni, V., Ye, J., Skiena, S., Wang, W.: Multi-view models for political ideology detection of news articles. In: Proceedings of the 2018 Conference on Empirical Methods in Natural Language Processing, pp. 3518–3527 (2018). https://doi.org/10.18653/v1/D18-1388

45. Le Pochat, V., Edelson, L., Goethem, T., Joosen, W., McCoy, D., Lauinger, T.: An audit of Facebook's political ad policy enforcement. In: Proceedings of the 31st USENIX Security Symposium (2022)

46. Lecun, Y., Bengio, Y., Hinton, G.: Deep learning. Nature **521**(7553), 436–444 (2015). https://doi.org/10.1038/nature14539

47. Leerssen, P., Ausloos, J., Zarouali, B., Helberger, N., Vreese, C.: Platform ad archives: promises and pitfalls. Internet Policy Rev. **8**(4) (2019). https://policyreview.info/articles/analysis/platform-ad-archives-promises-and-pitfalls

48. Leerssen, P., Dobber, T., Helberger, N., Vreese, C.: News from the ad archive: how journalists use the Facebook ad library to hold online advertising accountable. Inf. Commun. Soc. **0**(0), 1–20 (2021). https://doi.org/10.1080/1369118X.2021.2009002

49. Lindroos-Hovinheimo, S.: The proposed EU regulation on political advertising has good intentions, but too wide a scope (2022). https://europeanlawblog.eu/2022/02/23/the-proposed-eu-regulation-on-political-advertising-has-good-intentions-but-too-wide-a-scope/

50. Liu, B., Sheth, A., Weinsberg, U., Chandrashekar, J., Govindan, R.: AdReveal: improving transparency into online targeted advertising. In: Proceedings of the Twelfth ACM Workshop on Hot Topics in Networks, pp. 1–7 (2013). https://doi.org/10.1145/2535771.2535783

51. Lorenz-Spreen, P., Geers, M., Pachur, T., Hertwig, R., Lewandowsky, S., Herzog, S.: Boosting people's ability to detect microtargeted advertising. Sci. Rep. **11**(1), 15541 (2021). https://doi.org/10.1038/s41598-021-94796-z

52. Medienistitut, M.: Facebook: Datenmissbrauch um cambridge analytica — mainzer-medieninstitut.de

53. Medina Serrano, J., Papakyriakopoulos, O., Hegelich, S.: Exploring political ad libraries for online advertising transparency: lessons from Germany and the 2019 European elections. In: International Conference on Social Media and Society, pp. 111–121 (2020). https://doi.org/10.1145/3400806.3400820

54. Mehta, S., Erickson, K.: Can online political targeting be rendered transparent? Prospects for campaign oversight using the Facebook ad library. Internet Policy Rev. **11**(1), 1–31 (2022). https://doi.org/10.14763/2022.1.1648

55. Michener, G., Bersch, K.: Identifying transparency. Inf. Polity (2013). https://doi.org/10.3233/IP-130299

56. Mildner, T., Savino, G.L.: Ethical user interfaces: exploring the effects of dark patterns on Facebook. In: Extended Abstracts of the 2021 CHI Conference on Human Factors in Computing Systems, pp. 1–7 (2021). https://doi.org/10.1145/3411763.3451659

57. Holman, M.R., Schneider, M.C., Pondel, K.: Gender targeting in political advertisements (2015). http://journals.sagepub.com/doi/10.1177/1065912915605182

58. Mitchell, T.: Machine Learning. McGraw-Hill (1997)

59. Mittelstadt, B.: Automation, algorithms, and politics— auditing for transparency in content personalization systems. Int. J. Commun. **10**(0), 12 (2016)

60. Narayanan, A., Reisman, D.: The Princeton web transparency and accountability project. In: Cerquitelli, T., Quercia, D., Pasquale, F. (eds.) Transparent Data Mining for Big and Small Data. SBD, vol. 11, pp. 45–67. Springer, Cham (2017). https://doi.org/10.1007/978-3-319-54024-5_3

61. Nix, A.: The power of big data and psychographics — 2016 Concordia annual summit - YouTube (2016)
62. Novikova, I.: Big five (the five-factor model and the five-factor theory. Encycl. Cross-Cult. Psychol., 136–138 (2013). https://doi.org/10.1002/9781118339893.wbeccp054
63. de Oliveira, A.: Facebook advertisements and united states presidential campaigns (2019). https://www.cs.yale.edu/homes/jf/Oliveira610.pdf
64. Otterlo, M.: A machine learning view on profiling. In: Privacy Due Process and the Computational Turn: The Philosophy of Law Meets the Philosophy of Technology, pp. 41–64 (2013). https://doi.org/10.4324/9780203427644
65. Papakyriakopoulos, O., Hegelich, S., Shahrezaye, M., Serrano, J., König, P.: Social media and microtargeting: Political data processing and the consequences for Germany. Big Data Soc. 11(1), 87–110 (2018). https://doi.org/10.1515/spp-2019-0006
66. Risso, L.: Harvesting Your Soul? Cambridge Analytica and Brexit. Brexit Means Brexit (2018)
67. Ruohonen, J.: A dip into a deep well: online political advertisements, valence, and European electoral campaigning. In: van Duijn, M., Preuss, M., Spaiser, V., Takes, F., Verberne, S. (eds.) MISDOOM 2020. LNCS, vol. 12259, pp. 37–51. Springer, Cham (2020). https://doi.org/10.1007/978-3-030-61841-4_3
68. Savin, A.: The EU digital services act: towards a more responsible. Internet (SSRN Scholarly Paper (3786792)) (2021). https://papers.ssrn.com/abstract=3786792
69. Shen, F., Wu, H.D.: Effects of soft-money issue advertisements on candidate evaluation and voting preference: an exploration. Mass Commun. Soc. 5(4), 395–410 (2002). https://doi.org/10.1207/S15327825MCS0504_02
70. Serrano, J., Papakyriakopoulos, O., Shahrezaye, M., Hegelich, S.: The political dashboard: a tool for online political transparency. In: Proceedings of the International AAAI Conference on Web and Social Media, vol. 14, pp. 983–985 (2020)
71. Shiner, B.: General election 2019: unregulated digital political advertisements are damaging our democracy. https://www.democraticaudit.com/
72. Silva, M., Oliveira, L., Andreou, A., Melo, P., Goga, O., Benevenuto, F.: Facebook ads monitor: an independent auditing system for political ads on Facebook. In: Proceedings of The Web Conference 2020, pp. 224–234. Association for Computing Machinery (2020). https://doi.org/10.1145/3366423.3380109
73. Sneed, M.: The key to regulating Facebook and data collection companies is transparency. alb. LJ Sci. Tech. 30, 109 (2020)
74. Springer, A.: Making transparency clear, Los Angeles 5 (2019)
75. Springer, A., Whittaker, S.: Making Transparency Clear: The Dual Importance of Explainability and Auditability. IUI Workshops (2019)
76. Stabauer, M.: The impact of UI on privacy awareness. In: Nah, F.F.-H., Xiao, B.S. (eds.) HCIBGO 2018. LNCS, vol. 10923, pp. 513–525. Springer, Cham (2018). https://doi.org/10.1007/978-3-319-91716-0_41
77. Stewart, E.: Why everybody is freaking out about political ads on Facebook and google. Vox (2019). https://www.vox.com/recode/2019/11/27/20977988/google-facebook-political-ads-targeting-twitter-disinformation
78. Tene, O., Polenetsky, J.: To track or do not track: advancing transparency and individual control in online behavioral advertising. Minnesota J. Law Sci. Technol. 13, 281 (2012)
79. Venkatadri, G., Mislove, A., Gummadi, K.: Treads: transparency-enhancing ads. In: Proceedings of the 17th ACM Workshop on Hot Topics in Networks, pp. 169–175 (2018). https://doi.org/10.1145/3286062.3286089

80. Villegas, D., Mokaram, S., Aletras, N.: Analyzing online political advertisements. arXiv Preprint ArXiv:2105.04047
81. Witzleb, N., Paterson, M.: Micro-targeting in political campaigns. Political Promise and Democratic Risk (SSRN Scholarly Paper (3717561)) (2020). https://papers.ssrn.com/abstract=3717561
82. Wolfe, S.: Facebook approved 100 fake ad disclosures that were allegedly 'paid for' by every united states senator. Bus. Insider (2018). https://www.businessinsider.com/facebok-fake-ads-election-senators-2018-10
83. Wood, A.: Facilitating accountability for online political advertisements. Ohio State Technol. Law J. **16**, 520 (2020)
84. Zuiderveen Borgesius, F., et al.: Online political microtargeting: promises and threats for democracy. Utrecht Law Rev. **14**(1), 82–96 (2018)

Taking Responsibility: Who Should Be Held Responsible for the Misuse and Unintended Consequences of Social Media's AI Algorithms?

Ajith Sivakumar, Kennet Winter$^{(\boxtimes)}$, and Pradipa P. Rasidi

Department of Information Systems, Westfälische Wilhelms-Universität Münster, Münster, Germany
{ajith.sivakumar,k_wint10}@uni-muenster.de

1 Introduction

Ever since the Cambridge Analytica scandal in 2018, the role of social media companies in our public life has been under constant scrutiny. The everyday life of almost every member of our society is facilitated, shaped and influenced by the AI algorithms of social media companies. They shape public opinion by ranking and ordering each user's content and information [46].

With social media's influence over the public sphere, preventing the abuse of their algorithms, minimizing the possibilities of unintended uses, and mitigating the impact of unintended consequences for users and society becomes a vital task. The abuse of those algorithms can be done by the user or the social media company itself, willingly or unintendedly. Therefore, it is in society's best interest to prevent all abuse of social media's AI algorithms and force the companies to adapt quickly after misuses or harmful consequences become known.

This paper will discuss who must be responsible and what needs to be considered to prevent misuse and unintended consequences of social media AI. Mainly focusing on the question of who must be responsible for the next step, considering what actions were already taken by social media companies and governing bodies.

We will focus on discussing the cases of the United States and Indonesia, both of which are considered among the largest democracies globally. The United States ranks as the second-largest democracy, while Indonesia stands as the third largest. Moreover, both countries boast a significant number of social media users, with approximately 80.9% of the total population in the United States and 68.9% in Indonesia.

The selection of these countries for comparison stems from their shared experiences of witnessing the rise of social media algorithmic abuses and influence operations that have had significant implications on democratic decision-making. These issues have resulted in the reduction of civic spaces and the undermining of elections [3].

J. Bayer and C. Grimme (Eds.): AI, Human Rights and Ethics, LNAI 14400, pp. 63–76, 2025.
https://doi.org/10.1007/978-3-031-52082-2_5

Despite their commonalities in facing these challenges, the United States and Indonesia differ in their political experiences. The United States witnessed challenges in maintaining its democratic order since the rise of right-wing populism following the ascension of Donald Trump in 2016 [3]. On the other hand, Indonesia, with its more recent establishment of democratic institutions in 1998, has encountered predatory state practices whereby political-economic elites justify illiberal measures for resource extraction. This comparison between the United States and Indonesia allows us to examine the influence of social media algorithms on democratic decision-making within the context of two distinct regions: the Global North and the Global South.

The first chapter highlights how social media's algorithms influence the democratic decision-making process and gives examples of how these algorithms are currently misused. The second chapter introduces what actions have already been taken and what has been proposed by academic scholars. The third chapter will revisit the actions and double down on the question of which type of actor can be held responsible. Finally, the fourth chapter will wrap up with a discussion of the next steps.

1.1   Social Media's Algorithms Are Influencing the Democratic Decision-Making Process

Two of the most powerful internet platforms, Facebook and Google, are private companies that shape the everyday life of almost every member of our society.

They control which news are displayed and which information can be accessed by the user. Therefore, their algorithms shape the public debate by controlling which questions are discussed and with whom they are discussed, influencing collective decision-making in our democratic sphere [46].

Simons and Ghosh [46] describe the influence of Facebook and Google as "this infrastructure is not only critical to downstream economic activity, it influences the flow of ideas and information in our society, shaping how citizens discuss issues of common concern, organise to shape the world around them, and make a collective decision about fundamental matters of self-government."

This influence is caused by AI algorithms dynamically deciding which information is displayed to which individual user or set of users. Every time an algorithm displays any kind of information, a decision is made that effectively controls the access to information and news. Therefore, shaping the public debate [25].

However, algorithms are not sentient. They need to be developed, trained and evaluated. Furthermore, algorithms are political. Studies have proven that algorithms prioritise some social groups' interests over others [43]. Companies like Facebook and Google have no real incentives to change that because it furthers their primary goal to maximise their own profits. It has already been proven that tech giants like Google and Facebook do actually abuse their market power to engage in anti-competitive behaviour. There are almost no barriers to prevent their algorithms from being abused for their own benefit. The possibilities of abuse were clearly demonstrated during the US Congress antitrust hearings in July 2020 [31].

1.2 These Algorithms are Being Misused

Whether search engines or social media platforms, user interaction is essential to their business model. Therefore, AI-based algorithms increase the platform's service level, which might lead to biased information, echo chambers or filter bubbles [39]. Those phenomena can cause severe consequences.

Echo chambers describe the effect where social media limits the provided content to the user's perspective and leverages connecting to people with the same perspectives and opinions [8,14]. Filter bubbles are a similar phenomenon in which social media displays selected content regarding the user's preference. Research found that social media algorithms deepen ideological polarisation, leading to a loss of diversity of opinions and arguments as there are no other perspectives to confront [13,49]).

Combining such social media algorithms leverages an individual opinion towards a group opinion as there is less constructive public discussion. As a result, the group members are positively convinced regarding their own opinion and negatively view those with different opinions. This leads to a social division, which could be misused for political targeting, or the effective spread of political extremism or fake news [12,20,24]. The exposure of the Cambridge Analytica scandal in 2018 is an excellent example of political targeting. Personal data was used to categorise voters into specific groups and to positively influence them with targeted political advertising during the election campaign of 2016 [4,7,50].

2 What Actions Have Already Been Taken?

This chapter will briefly introduce actions that have already been implemented and ideas that researchers have proposed to fight the misuse and unintended consequences of social media's AI algorithms. While only introducing and explaining the possible actions in this chapter, the upcoming chapter will discuss the possibilities, effectiveness and implications of the mentioned actions.

In order to analyse the question of who must be responsible for the next major step, we identified three potential actors: the users of social media platforms, the corporations operating the social media platforms, and governmental bodies on the federal and international levels.

2.1 Addressing the End-Users of Social Media Platforms

2.1.1 Education in Digital Literacy

In Indonesia and the United States, digital literacy has been the main course of action to fight against misinformation and filter bubbles. For example, Indonesian and American governments, academics, and civil society organisations spearheaded digital literacy programs in schools, universities, ministries, companies, and neighbourhood communities.

Scholars have conducted research projects assessing the level of digital literacy and education to measure citizens' resilience in the face of disinformation and filter bubbles [5, 26, 29, 37]. The model relies on the premise that the more literate someone is, the less they may believe in disinformation and be trapped in filter bubbles.

2.2   Addressing the Corporations Operating Social Media Platforms

2.2.1   Preventing Misuse by Algorithm Design

A survey in 2022 from the Reuters Institute ascertained that 82% of respondents from Kenya admitted using social media as a news source. By comparison, more than 50% of adults in Portugal, Poland, Romania, Hungary, Bulgaria, Slovakia, and Croatia stated that they consume their news on social media [23]. As we have already identified, the combination of social media algorithms and fake news risks users. Social media consumption can be considered a double-edged sword. Even if the news is low cost and easily accessible, it can be of poor quality or even be fake news and could be wrongly perceived as truthful by users [45]. Therefore, it is essential to know to what extent social media platforms counteract threads caused by their algorithms, such as fake news and political targeting.

The first step of social media is the reporting function as Facebook does. It is possible to report posts that contain or spread fake news. In addition, social media like Twitter, YouTube and Facebook work together with fact-checkers [11]. Facebook cooperates with news agencies such as ABC, DPA, and the research network Correctiv to identify fake news, illustrate to users through images or symbols that the content may not correspond to the facts, and offer them the opportunity to obtain further information about the post. Facebook deletes fake news that can lead to life-threatening situations; according to the company, 25 million pieces of such content were removed from January to March 2021 [2, 41].

Active methods to combat fake news are the generic display of correct information in a thread with a large amount of fake news, e.g., feeds related to Covid-19. Instagram often displayed direct links to legitimate sources such as the Federal Ministry of Health and, as Facebook did, reduced user interaction after the content was recognised as fake news [11, 28]. YouTube, for example, started demonetising feeds of content related to the war between Russia and Ukraine to reduce the financial incentive to spread fake news [51].

2.3   Addressing Governmental Bodies on Federal and International Levels

2.3.1   Regulations and Antitrust

Within the last years, multiple antitrust cases were opened, i.e., by the US congress [31], the EU parliament, and the German antitrust agency [6], to regulate big tech companies. Antitrust is a legal tool to address market power and anti-competitive behaviour alongside other classical tools like lawsuits, prohibition of mergers, breaking up monopolies, or introduction of competition [46].

However, those antitrust cases were primarily reactive and aimed at changing the status quo instead of actively shaping how Google, Facebook, and its algorithms are influencing society in the future. As a result, using antitrust cases is a slow and reactive regulatory tool. It is not fit for purpose and cannot sufficiently adapt quickly enough to regulate fast-changing AI algorithms in a digital environment.

2.3.2 Fighting Misinformation with Active Policing

In Southeast Asia, governments have been responding to misinformation by persecuting its dissemination as a criminal offence [32,47,48]. Misinformation is often seen as a security threat and thus is approached by security measures as a threat that endangers "national stability". This term is used arbitrarily and vaguely, which may include anyone perceived as threatening the government, including democratic political opposition. With national laws in place, police forces are actively deployed to persecute people accused of spreading misinformation.

In Indonesia, for example, tackling misinformation is often conflated with tackling the rise of populist Islamist movements seen as opponents of the incumbent government [9,19]. The law on "spreading hoax" is selectively enforced on the people considered populist Islamists-which by itself can be criminalised if the activity is considered against the state ideology Pancasila-while misinformation benefiting the government (misleading information on political opposition) remains untouched. The Indonesian Ministry of Communication and Informatics publishes regular "stamps" determining whether a piece of information is considered a "hoax" or not, and police forces act on those "stamps".

In 2020, after the Indonesian government saw an increase in protests from civil society organisations and human rights activists, police forces were also deployed to arrest activists based on the accusation that they were spreading misinformation against the government [44].

Only in these recent years have scholars started noticing that this may lead to an increase in authoritarian practices. Previously, scholars have primarily supported the government narrative that policing misinformation to deter Islamic populists is necessary, echoing the American narrative that there is a "firehose of falsehood" undermining the government [1,17,22,26,40,42]. There was a huge blind spot on the possible undemocratic policing in that effort.

3   Who Must Be Responsible for the Next Steps?

Having introduced some actions in the previous chapter, this chapter will discuss their pros and cons, introduce new ideas, and examine the environment needed to hold the different actors responsible for the misuse and unintended consequences of social media's AI algorithms.

3.1 Responsibilities of the End-Users of Social Media Platform

The following sub-chapter highlights that focusing on digital literacy alone is not practical. However, the second sub-chapter emphasises that filter bubbles are co-created by the end-users and their individual use of social media.

3.1.1 Digital Literacy is not Actually Effective

Despite the large number of efforts conducted by government and civil society organisations across the United States and Indonesia, there remains a question of whether digital literacy is an effective means to tackle the issue posed by algorithmic unaccountability and the misinformation problem that followed.

Here we argue that digital literacy cannot be the panacea as usually proposed by such educational programs. The biggest issue is that such digital literacy programs barely consider users' media practices. Participants are gathered in workshops, training, and focus group discussions with programmatic modules at hand [5]. An interview with an Indonesian activist reveals that such programs are often conducted sporadically with no reliable measurement to gauge their impact appropriately, while critics noted that the top-down, patronising conduct of such programs is measured only by the number of participants who attended [9,16]. A study one of us conducted [38] on Twitter influencers in Indonesia shows that even digitally literate users may still spread misinformation.

In the case of Indonesian Twitter influences discussed in the research cited above, Twitter's hashtag and the easiness of going viral induce a feeling of belonging to a collective of concerned citizens. Therefore, creating a community of practice. At the same time, the situation in Indonesia, where political elites own hyperpartisan media, has given way for people to construe the act of spreading misinformation as meaningful participation in democratic elections. This situation is fuelled further by the culture of secrecy, conspiracy theory, and the lack of meaningful political participation methods for ordinary citizens.

3.1.2 Personal Echo Chambers are Co-created by the Individual User

While plenty of scholars have written about the filter bubble and how it normalises the echo chamber, not enough attention has been paid to the other side of the coin: how users engage with the algorithms themselves. A few studies [15,27] have hinted that, beyond the technological deterministic assumption of filter bubbles, users do play a part in creating their own echo chamber by playfully interacting with algorithms. First, they dislike YouTube videos they do not favour, hoping to see less similar content. Furthermore, they frequently retweeted and replied to users with similar interests to them as means to incentivise algorithms to bring those users and others like them to their timeline. There is an intentionality to cater their feeds to their own liking. The same studies also show that users deliberately find news outlets with different perspectives, technically breaking out of the filter bubble social media supposedly created.

This approach is interesting to follow. A participation-observation fieldwork research that one of the authors of the paper conducted when examining the phenomenon of filter bubble in Jakarta during the 2017 election [38] shows that, when dealing with political uncertainty, users would deliberately curate and distribute content that we typically designate as misinformation. While some research [18] has argued that cognitive bias may influence people's decision to disseminate misinformation, this was not the case in the research Rasidi conducted. It was not bias that drove the action but a desire to uncover 'truths' perceived as hidden behind electoral dramaturgy. His interlocutors intentionally tweaked their social media feed to show information regarding the election and distributed rumours, gossips, hearsays, and supposed "live tweeting" to shed light on the uncertain political situation at the time.

Not all social media platforms were made equal. For interlocutors in that research, Twitter was considered the only appropriate platform to gossip about politics and uncover hidden truths. They do not do something similar on Instagram and Facebook, as those platforms were considered unsuitable for political conversation due to their connection to relatives and families. Therefore, the users' media practices must be considered when discussing echo chambers. Echo chambers are co-created by algorithmic ordering and the user's own decisions in exercising their informational or political agency.

3.1.3   Call for a More Holistic Approach

Making the end-user responsible for identifying misinformation cannot be the sole course of action. However, digital literacy is essential, and education must continue. Furthermore, without action from governments and platform providers, the end-user alone will be unable to stop all arising consequences and misuses of social media's AI algorithms.

3.2   Responsibilities of the Corporations Operating Social Media Platforms

After identifying the need for a more holistic approach in the last chapter, this section discusses the measures currently taken by social media providers and whether their actions can be trusted within the current capitalist society.

3.2.1   Dealing with Spam: The Current Measures Must Be Expanded

The current measures of social media, such as Facebook and Instagram, have identified 1.8 billion spam content in the first quarter of 2022 [36]. Even after 1.6 billion fake accounts have already been deleted in the first quarter of 2022 [35].

Facebook defines spam as "a broad term used to describe content that is designed to be shared in deceptive and annoying ways or attempts to mislead users to drive engagement" [36]. Content, such as posts, photos and videos, is counted individually. This means that if, for example, a Facebook post with four photos is removed, a total of five pieces of spam content is counted as detected. Considering that the reporting function detected only 0.3% of this detected spam content, it becomes clear that active fact-checking is the most effective approach to counteract misleading content. Moreover, this list does not consider the propagation speed and the impact on the number of users. One problem of this manual approach is scalability. For example, the number of monthly active Facebook users in the second quarter of 2022 is 2.93 billion users [34]. In a fictitious calculation in which every second active user posts a message, the extrapolation in one quarter (90 days) amounts to 131.85 billion feeds.

Due to the volume and velocity of data produced by social media platforms, the current measures of social platforms such as Facebook and Instagram are insufficient to verify all posts on time. This is why Facebook, for example, is actively trying to find additional methods to combat the spread of spam content [36].

3.2.2 Dealing with Filter Bubbles and Fake News

One approach to combat the drawbacks of AI algorithms could be to alter them with minimal effort. For example, filter bubbles and echo chambers could be relieved by displaying random news from public news channels and prioritising high-quality verified content [21].

Another approach to combat political fake news would be to increase cooperation with the government and publish accurate, high-quality content or a cross-reference to lawful government pages, such as in the context of Covid-19. However, it is also important to note that this approach could be contraindicated in the case of illiberal governments that themselves practice media control and propaganda.

The fact that illiberal governments are able to take advantage of their power, does not render government intervention generally useless. Regulation can set limits on corporations and hold them accountable for the provided services. Facebook is a prime example to demonstrate the profit-driven business model of such collaborations.

3.2.3 Private Corporations in a Capitalist Society

Living in a capitalist society, social media companies have no monetary incentive to fight the misuse of their algorithms, because they reap the financial benefits of spreading misinformation and disinformation on their platforms [30]. Therefore, without legal regulations, combined with pressure from society and the media, there might be no incentive at all to fight misinformation.

The authors of this paper conclude that, first and foremost, the companies are responsible for the AI algorithms they need to address misuses and problems on their platforms. However, if the environment they operate in does not change, private corporations have no reason to act on their responsibilities. Therefore, it is the responsibility of federal and international governments to create regulations and the responsibility of society to create public pressure to change that environment.

3.3 Responsibilities of Governmental Bodies

After identifying the need for governmental bodies to force private corporations to act in the interest of society in the previous section, this sub-chapter advises against active policing of social media platforms, and it proposes the idea of mandatory ethical committees overseeing the company-internal commissions.

3.3.1 Active Policing as an Undemocratic Measure

As scholars and activists in Indonesia have later realized, policing misinformation has resulted in undemocratic arrests and cracking down against dissents. Active policing is not a solution. Following scholars' concerns about the rise of digital authoritarianism [47], regional civil society organisations in Southeast Asia and Asia Pacific, such as SafeNet and EngageMedia, have eventually addressed the failings of active policing.

3.3.2 Utilising an Ethics Committee to Oversee and Evaluate Algorithmic Impact

Ethics Committees in Medical Research

In medical research, ethics committees evaluate the ethical and legal legitimacy of medical research projects on humans (and animals). These committees exist on different levels. For example, in Europe, there are local and federal ethics committees and a European committee overseeing the federal committees. An experiment will only be allowed or funded by universities and research institutes if it passes a screening by an ethical commission as long as it concerns the well-being of humans or animals. The ethical commission will evaluate the experiment's dangers, harm, and possible impact and put regulations for the experiment in place.

Ethics Committees Within Social Media Companies

This practice could be adapted for AI algorithms. Before being deployed into production, the algorithm must be evaluated by an ethics commission that will evaluate the algorithmic impact on the general population and possibly affected minorities. However, this is not a new idea. Several well-known social media companies already have an ethical team in place [33].

In the light of the findings of Sect. 3.2.3, the validity, effectiveness, and legitimacy of company-internal ethical committees must be doubted. Without an active and legally binding supervision, society cannot know whether these company-internal committees are a tool for the companies against public back-lash or whether they can actually fulfil their role as an ethical oversight.

The appointment of an independent committee comprising stakeholders from civil society, academia, and private companies is crucial for effective oversight. In Indonesia, a trial initiative is underway to establish an ethical committee responsible for overseeing social media companies. This committee involves col-laboration between civil society actors and digital marketing agencies, aiming to promote accountability and transparency among social media platforms. How-ever, the impact of this committee remains to be seen due to the lack of pressure and consequences imposed on social media companies.

Call for Governmental Oversight

The German Federal Data Ethics Commission "believes that the State has a particular responsibility to develop and enforce ethical benchmarks for the digital sphere that reflect [their] value system" [10]. In order to do so, the State or an alliance of states needs to "act from a position of political and economic strength on a global stage" against "private corporations that are, for the most part, exempt from democratic legitimacy and oversight".

The authors of this paper endorse this position and call for mandatory and external ethics committees actively overseeing the development, deployment, use and misuse of AI algorithms that impact the democratic decision-making process to a great extent. However, individual national legislation to enforce these committees is not enough. International legislation that reflects different value systems is needed to support governments without enough economic or political power to oppose corporations like Facebook or Google.

4   Conclusion

Social media's AI algorithms are influencing the public sphere. Therefore, it is crucial to prevent the abuse of their algorithms, minimise the possibilities of unintended uses, and mitigate the impact of unintended consequences for users and society. Abuse of those algorithms can be done by users or the social media companies themselves. Therefore, it is in society's best interest to prevent all abuse of social media's AI algorithms and to hold the companies responsible for adapting quickly after misuses or harmful consequences become known.

This paper discussed who must be responsible and what needs to be consid-ered to prevent misuse and unintended consequences of social media AI. Mainly focusing on the question of who must be responsible for the next step, consider-ing what actions were already taken by social media companies and governing bodies.

In the first chapter, we highlighted the importance of social media's algo-rithms, their impact on democratic decision-making processes, and examples of the misuse of those algorithms. In the second chapter, we identified three differ-ent levels of actors: the end-users, social media providers and governing bodies.

We also introduced what actions have already been taken or considered to minimise the algorithm's abuse. Finally, in the third chapter, we analysed the previously introduced actions. Furthermore, we discussed all three of the mentioned levels of actors.

When discussing the three levels of actors, we found that a holistic approach is necessary. Simply focusing on one of the three layers will not solve societies' problems by misusing social media's AI algorithms. The end-users are co-creating echo chambers by using social media platforms in a specific way. Social media companies are profiting financially from the spread of misinformation on their platform. Without public pressure from society and regulations from governing bodies, there is no incentive for social media companies to be accountable. Therefore, regulations are needed to set a legal framework in which social media companies are forced to operate. To do so, the government must be in a political and economic position of power or cooperate with different governments. Unfortunately, enforcing those regulations can also lead to situations that are harmful to the democratic society.

In order to create a regulatory framework for the social media companies to work in, an international ethics committee overseeing the mandatory ethical committees within private corporations has been proposed. This ethics committee needs to consider the different value systems of the participating countries.

In future research, the following questions need to be addressed: How are the members of such an ethical committee appointed? Who do they represent and who do they answer to? And finally, how much authority and power are governments willing to delegate to an ethical committee, and how can they be democratically controlled?

However, the constraints of this paper did not allow endeavouring to explore this path. Therefore, further discussions and research are recommended, to evaluate whether such an overseeing ethics committee would be a suitable means to change the rules of social media providers.

References

1. Akmaliah, W.: Kebenaran yang terbelah: Populisme islam dan disinformasi politik elektoral. MAARIF **14** (2019). https://doi.org/10.47651/mrf.v14i1.53
2. Ballweg, S.: Mit faktencheck gegen fake news. Deutschlandfunk (2017). https://www.deutschlandfunk.de/soziale-medien-mit-faktencheck-gegen-fake-news.724.de.html?dram:article_id=378182
3. Bennett, W.L., Livingston, S.: The disinformation order: disruptive communication and the decline of democratic institutions. Eur. J. Commun. **33**(2), 122–139 (2018). https://doi.org/10.1177/0267323118760317
4. Berghel, H.: Malice domestic: the Cambridge analytica dystopia. Computer **51** (2018). https://doi.org/10.1109/MC.2018.2381135
5. Bu, D.: Kerangka literasi digital indonesia (2017). https://literasidigital.id/books/kerangka-literasi-digital-indonesia/
6. Bundeskartelamt: Facebook, exploitative business terms pursuant to section 19(1) GWB for inadequate data processing (2019). https://www.bundeskartellamt.de/SharedDocs/Entscheidung/EN/Fallberichte/Missbrauchsaufsicht/2019/B6-22-16

7. Cadwalladr, C., Graham-Harrison, E.: Revealed: 50 million Facebook profiles harvested for Cambridge analytica in major data breach — news — the guardian. The Guardian (2018)
8. Cinelli, M., de Francisci Morales, G., Galeazzi, A., Quattrociocchi, W., Starnini, M.: The echo chamber effect on social media. Proc. Natl. Acad. Sci. USA **118** (2021). https://doi.org/10.1073/pnas.2023301118
9. Citra, D., Rasidi, P.P.: In the name of religious harmony: challenges in advancing religious freedom in digital Indonesia (2022)
10. German Data Ethics Commission: Opinion of the data ethics commission (2019). https://www.bmj.de/SharedDocs/Downloads/DE/Themen/Fokusthemen/Gutachten_DEK_EN.pdf?_blob=publicationFile&v=2
11. Cotter, K., DeCook, J.R., Kanthawala, S.: Fact-checking the crisis: Covid-19, infodemics, and the platformization of truth. Soc. Media Soc. **8** (2022). https://doi.org/10.1177/20563051211069048
12. DiFranzo, D., Gloria-Garcia, K.: Filter bubbles and fake news. XRDS: Crossroads, The ACM Magazine for Students, vol. 23 (2017). https://doi.org/10.1145/3055153
13. Du, S., Gregory, S.: The echo chamber effect in twitter: does community polarization increase? Studies in Computational Intelligence, vol. 693 (2017). https://doi.org/10.1007/978-3-319-50901-3
14. Dubois, E., Blank, G.: The echo chamber is overstated: the moderating effect of political interest and diverse media. Inf. Commun. Soc. **21** (2018). https://doi.org/10.1080/1369118X.2018.1428656
15. Dutton, W.H., Reisdorf, B.C., Dubois, E., Blank, G.: Social shaping of the politics of internet search and networking: moving beyond filter bubbles, echo chambers, and fake news. SSRN Electron. J. (2017). https://doi.org/10.2139/ssrn.2944191
16. Eps, V.: Advancing data justice research and practice. EngageMedia (2022)
17. Fauzi, A.I.: Nationalism and Islamic populism in Indonesia. Henrich Boll Stiftung Southeast Asia (2018)
18. Fernbach, P.M., Boven, L.V.: False polarization: cognitive mechanisms and potential solutions. Curr. Opin. Psychol. **43** (2022). https://doi.org/10.1016/j.copsyc.2021.06.005
19. George, C.: Hate spin: the manufacture of religious offense and its threat to democracy (2016). https://doi.org/10.1177/0267323118824881
20. Grimes, D.R.: Online bubbles. The Guardian (2017). https://www.theguardian.com/science/blog/2017/dec/04/echo-chambers-are-dangerous-we-must-try-to-break-free-of-our-online-bubbles
21. Howard, P.: Is social media killing democracy? Oxford Internet Institute (2016)
22. Hui, J.Y.: Social media and the 2019 Indonesian elections: hoax takes the centre stage. Southeast Asian Affairs 2020 (2020). https://doi.org/10.1355/9789814881319-010
23. Share of adults who use social media as a source of news in selected countries worldwide as of February 2022 (2022). https://www.statista.com/statistics/718019/social-media-news-source/
24. Kelly, M.: Political polarization and its echo chambers: surprising new, cross-disciplinary perspectives from Princeton (2021). https://www.princeton.edu/news/2021/12/09/political-polarization-and-its-echo-chambers-surprising-new-cross-disciplinary
25. Klonick, K.: The new governors: the people, rules, and processes governing online speech. Harvard Law Rev. **131** (2018)
26. Kusman, A.P.: Hoaxes and fake news: a cancer on Indonesian democracy (2017)

27. Lim, M.: Freedom to hate: social media, algorithmic enclaves, and the rise of tribal nationalism in Indonesia. Crit. Asian Stud. **49** (2017). https://doi.org/10.1080/14672715.2017.1341188

28. Maurer, L.G.: Cyber-silencing the community: Youtube, divino group, and reimagining section 230. Washington J. Law Technol. Arts **17**(2), 172 (2022)

29. Moore, R.C., Hancock, J.T.: A digital media literacy intervention for older adults improves resilience to fake news. Sci. Rep. **12** (2022). https://doi.org/10.1038/s41598-022-08437-0

30. Mosseri, A.: Working to stop misinformation and false news (2017). https://www.facebook.com/formedia/blog/working-to-stop-misinformation-and-false-news

31. Nadler, J., Ciciline, D.N.: Investigation of competition in digital markets. majority staff report and recommendations. U.S. House of Representatives. Subcommittee on Antitrust, Commercial and Administrative Law of the Committee on the Judiciary (2020)

32. Neo, R.: The securitisation of fake news in Singapore. Int. Polit. **57** (2020). https://doi.org/10.1057/s41311-019-00198-4

33. Novet, J.: Facebook forms a special ethics team to prevent bias in its A.I. software (2018). https://www.cnbc.com/2018/05/03/facebook-ethics-team-prevents-bias-in-ai-software.html

34. Anzahl der monatlich aktiven facebook nutzer weltweit vom 1. quartal 2009 bis zum 2. quartal 2022 (2022). https://de.statista.com/statistik/daten/studie/37545/umfrage/anzahl-der-aktiven-nutzer-von-facebook/

35. Fake accounts (2022). https://transparency.fb.com/data/community-standards-enforcement/fake-accounts/facebook/

36. Spam (2022). https://transparency.fb.com/data/community-standards-enforcement/spam/facebook

37. Polizzi, G.: Information literacy in the digital age: why critical digital literacy matters for democracy. Informed Societies (2020). https://doi.org/10.29085/9781783303922.003

38. Rasidi, P.P.: Melampaui 'literasi digital': Kelindan teknokapitalisme, ideologi media, dan hiperhermeneutika. Jurnal Komunikatif (2021)

39. Rhodes, S.C.: Filter bubbles, echo chambers, and fake news: how social media conditions individuals to be less critical of political misinformation. Polit. Commun. **39** (2022). https://doi.org/10.1080/10584609.2021.1910887

40. Riyanto, G.: Media sosial, habitat alami populisme religius? MAARIF **12**(1), 87–105 (2017)

41. Ruppert, J.: Was social-media-plattformen gegen falschmeldungen... (2021). https://www.br.de/nachrichten/netzwelt/was-social-media-plattformen-gegen-falschmeldungen-tun,SgAGfv1

42. Salma, A.N.: Defining digital literacy in the age of computational propaganda and hate spin politics. KnE Social Sciences (2019). https://doi.org/10.18502/kss.v3i20.4945

43. Sankin, A.: How activists of color lose battles against Facebook's moderator army. Reveal (2017). https://revealnews.org/article/how-activists-of-color-lose-battles-against-facebooks-moderator-army/

44. Sastramidjaja, Y., Rasidi, P.P.: The hashtag battle over Indonesia's omnibus law: from digital resistance to cyber-control. ISEAS Perspective (2021)

45. Shu, K., Sliva, A., Wang, S., Tang, J., Liu, H.: Fake news detection on social media. ACM SIGKDD Explor. Newslett. **19** (2017). https://doi.org/10.1145/3137597.3137600

46. Simons, J., Ghosh, D.: Utilities for democracy: why and how the algorithmic infrastructure of Facebook and google must be regulated. Foreign Policy at Brookings (2020)
47. Sinpeng, A., Tapsell, R.: 1. from grassroots activism to disinformation: social media trends in southeast Asia. From Grassroots Activism to Disinformation (2021). https://doi.org/10.1355/9789814951036-002
48. Sombatpoonsiri, J.: Ecuritizing "fake news": policy responses to disinformation in Thailand. Grassroots Activism to Disinformation: A Selection, pp. 105–125 (2020)
49. Spohr, D.: Fake news and ideological polarization. Bus. Inf. Rev. **34** (2017). https://doi.org/10.1177/0266382117722446
50. Wong, A.: Britische datenschützer verhängen höchststrafe gegen facebook (2018). https://www.zeit.de/digital/datenschutz/2018-10/cambridge-analytica-datenskandal-facebook-geldstrafe
51. YouTube: Richtlinien für werbefreundliche inhalte (2022). https://support.google.com/youtube/answer/6162278

Social Bots Spreading Disinformation About Finance: Research Trends, and Ethical Challenges

Janina Pohl[1]([✉])(ID), Marie Griesbach[1], Alireza Samiei[2], and Adelson de Araujo[3](ID)

[1] Department of Information Systems, Westfälische Wilhelms-Universität Münster, Münster, Germany
{janina.pohl,mgriesba}@uni-muenster.de
[2] The Leiden Institute of Advanced Computer Science, Universiteit Leiden, Leiden, The Netherlands
a.samiei@umail.leidenuniv.nl
[3] Faculty of Behavioural, Management and Social sciences, University of Twente, Enschede, The Netherlands
a.dearaujo@utwente.nl

Abstract. The ever-rising danger of social bots spreading disinformation about stock markets and the coordinated usage of mass misinformation tactics in social media can create conditions for malicious groups to manipulate market values and investor behaviors in weak positions. Several studies have proposed approaches to dealing with disinformation escalated by modern technologies, and regulatory bodies need to absorb the lessons given. In this paper, we will discuss the problem of disinformation in finance and the role of social bots. To have a birds-eye view of the most prominent topics in the academic literature, we conduct a topic modeling analysis to depict broad patterns, e.g., the connectivity between subjects and their topicality. Subsequently, we provide some thoughts and directions on what new regulations might require considering to weigh the benefits and drawbacks of regulatory actions. As a result, we consider a number of broad and interdisciplinary implications of financial and ethical regulations.

1 Introduction

Artificial Intelligence (AI) accelerates many processes and procedures in every part of modern life and confronts society and business with new challenges. For example, autonomous trading agents can manage stocks at an impressive scale in finance. At the same time, social media platforms facilitate sharing information about financial assets, e.g., stocks and cryptocurrencies [33]. Studies have shown a positive relationship between stocks' relevance on Twitter and their importance in the market [8,32]. Nevertheless, the information found on social media is not necessarily accurate or trustworthy, and malicious groups can use AI

J. Bayer and C. Grimme (Eds.): AI, Human Rights and Ethics, LNAI 14400, pp. 77–95, 2025.
https://doi.org/10.1007/978-3-031-52082-2_6

technology to scale the spreading of disinformation and profit from social media manipulation. AI-powered social bots can spread disinformation with potentially harmful consequences since an increasing number of people get their news from social media [18,53]. In their work, Grimme et al. (2017) define social bots as "superordinate concept, which summarizes different types of (semi-) automated agents. These agents are designed to fulfil a specific purpose by means of one- or many-sided communication in online media". Hence, social bots are programmed to spread a specific message, e.g., about a stock, by sending it to others to promote their operator's opinion [17].

In research, social bots and disinformation are relatively recent subjects of discussion. In this paper, we conducted a systematic topic modelling analysis with papers published from 2016 to 2021 to find which topics are currently present, how they connect, and how frequently they were addressed over time. As we will show, social bots and disinformation started to be discussed beyond simple campaign promotion strategies in 2018, and other literary topics arose even more recently, e.g., cryptocurrencies' "pumps and dumps" after 2020. Our topic modelling analysis aims to provide a bird's-eye view of the scientific literature about this subject.

After identifying the lines of scientific investigation, we explore some ethical, methodological, and regulatory issues in the intersection of finance, disinformation, and AI-powered social bots. For instance, how to decide whether a bot informs or manipulates? Should authorities pass laws to reduce disinformation in social networks? How to fight disinformation without restraining freedom of speech? Our work aims to elaborate on these ethical questions and provide some arguments on the requirements of regulations.

In Sect. 2.1, we will first discuss the influence of disinformation on the financial market, the mechanisms of decision making, and how they can be exploited by spreading disinformation. In Sect. 2.2, we describe our topic modelling approach. Since social bots are a suitable and often used tool for spreading disinformation, in Sect. 2.3 we will investigate the current state of research in more detail. Finally, in Sect. 3, the results will be discussed concerning ethical challenges and possible regulations in that field, followed by a short conclusion and outlook in Sect. 4.

2   Background

2.1   Financial Markets and Disinformation

Investors may use social media to obtain and share stock market information, which also puts them at risk of being exposed to misleading information about markets [42]. As Petratos (2021) argues [39], misleading information, including disinformation and fake news, has been a critical issue in the past decade, with examples of the interference in the 2016 U.S. presidential elections and the spread of false information about the COVID-19 pandemic during the past two years, the threat of misleading information has been becoming an urgent issue for business practices. Furthermore, Petratos (2021) discusses that due to technological advancements and the adaptation of financial services to digital technologies, the demand for online financial services has witnessed a steady upward trend during the pandemic [39]. Since modern financial markets are highly dependent on information flows and communication systems, this interconnected nature of new financial marketplaces makes them especially vulnerable to the spread of inaccurate information [29].

Most social bots' activities happen on social media platforms; therefore, retail investors who are the main participants on social media platforms are the primary target of manipulation in terms of spreading disinformation using social bots [23]. It does not mean that institutional investors are entirely immune to the danger of disinformation with the use of social media bots, but the scope of this research is more limited to the effect of disinformation on individual investors.

The U.S. Securities and Exchange Commission (SEC) issued an investors' alert in November 2015 concerning the rising danger of using social media platforms to perform market manipulation: "One-way fraudsters may exploit social media is to engage in market manipulation, such as spreading false and misleading information about a company to affect the stock's share price. Wrongdoers may perpetuate stock rumours on social media, as well as on online bulletin boards and in Internet chat rooms" [42,51]. One of the examples of market manipulation techniques mentioned was the pump-dump schemes. Renault (2017) defines the pump-dump schemes as "touting a company's stock through false or misleading statements in the marketplace to inflate artificially (pump) the price of a stock. Once fraudsters stop hyping the stock and sell their shares (dump), the price typically falls." He showed that pump-and-dump schemes mainly target small-capitalization stocks with low liquidity traded in the over-the-counter (OTC) markets using various forms of spreading false information, including social media platforms.

Another set of threads comes from advances in artificial intelligence regarding computer vision and audio. For example, the term "deep fakes" has grown considerably in popularity recently due to its risky capabilities of dissuading people with artificially generated images or videos, e.g., the "Synthesizing Obama" program [46]. More recently, text-based methods are used in the same sense to create plausible deep fake messages [11]. With today's accessibility for engineers to work with more complex AI models, these deep fakes can be in the hands of groups of malicious financial agents manipulating investors in weak positions.

2.1.1   The Impact of Disinformation

Besides the direct impact of fraudulent news on financial markets, a study performed by Kogan et al. (2018) discussed whether misinformation campaigns have a broader and more indirect effect on the investor's behaviour. After a shock to investors' trust in news caused by an SEC's announcement in 2014 about an ongoing investment on shared financial news networks, they examined how the trust of the retail investors was affected. A key finding of the mentioned study was that after investors realized the liability of information they received from socially shared financial news networks, they discounted real news from the mentioned platforms but lost their trust in even real news coming from these platforms. Furthermore, their results show a 55% increase in trading volume over three days after releasing fake news articles [23]. This finding is of vital importance since it shows that if the spread of disinformation using social bots becomes more widespread in the future, it will seriously affect the stability and trends of financial markets and the conduct and trust of especially retail investors in the future. Another research by Clarke et al. (2021) explored how investors' attention is affected by fake news articles and how the stock market reacts to disinformation. On average, fake news articles attract 83.4% more page views than original news articles. Furthermore, previous studies show that fake news articles are 70% more likely to be shared on social media platforms such as Twitter [3,55].

The financial market is affected by social media and also by social bots. There is a positive correlation between stock capitalization and discussion volume. Therefore, spreading information on social media could considerably influence the stock's financial importance [47]. The Wall Street Journal publishes the IHS U.S. sentiment index, which measures social media sentiment using a deep learning architecture. Every minute, ten percent of all tweets are analysed and weighted. Each tweet is assigned a rating based on the pre-trained model (124,6 million training data) and even uses emojis to predict the sentiment [28]. Therefore, social bots can have a direct influence on decision-making. Social engineering, troll farms, or social bots aim to spread disinformation. Besides these covert operations, disinformation can also be outspread via public media or websites [40].

Before taking a closer look at sentiment and the effects of disinformation, it is essential to distinguish between disinformation and misinformation. Misinformation is defined as unintentionally mistaken content. It includes false information such as inaccurate statistics, translations, or satire taken seriously. On the other hand, disinformation is defined as fabricated or deliberately manipulated content. The main difference is that it is intentionally created to spread conspiracy theories or rumours. Another even more potent type of harmful information is malinformation. It describes the deliberate publication of information that was not meant to be published. Publishing is motivated by personal or corporate interest, and the content's context, date, or the time has been modified [57]. When discussing social bots, the focus is on disinformation. Social bots are used to influence the sentiment, which is analysed to predict future trends [16].

Investors could be influenced without noticing that they are faced with fake news or even aware that the information is misleading. However, even if they are aware of that, they still may be influenced. This phenomenon is called the continued influence effect, which describes that the actions or decisions of people are still influenced by incorrect information even if they know that it is incorrect - knowing that a piece of information is incorrect leads to updating the inferences but does not lead to ignoring the information as a whole. In order to tackle the issue of fake news, traders could be sensitized to fake news beforehand. This leads to a trade-off between being more sceptical towards news and recognizing misleading information and the risk of missing out on important news [16]. Another potential threat for investors is the limited attention span bias, as attention is determined by boredom and flow [58]. The scarce resource of attention influences behaviour and decisions. Boredom describes that the appearance of many opportunities is expected to lead to diverting previous attention. The other factor is the flow, which is keeping attention if other opportunities are sparse [58]. Summing up, human decision-making is influenced by the difficulty of recognizing fake news, the ability to make rational decisions, and the limit of attention.

2.1.2 Fully Automated Processes

However, what if the whole process of analysing social media and decision-making is automated? Social bots can influence the sentiment predicted via a Natural Language Generation algorithm. This is the first step where manipulation can occur and, therefore, be prevented. Compared to political social bots, the bots used in the finance sector are still simple. Tardelli et al. [49] claim that the finance social bots are that simple because they were not hunted yet. So, it can be expected that those bots will get more sophisticated in the future.

The sentiment analysis result, besides other information, is used for decision-making. A trading bot has three options: (1) buy a stock, (2) sell a stock, (3) do nothing. Nan et al. [36] compare the performance of trading bots for Microsoft with and without using social media sentiment analysis. They came up with the result that trading bots using the sentiment analysis performed better and therefore generated the highest profit compared to the other bot or random policy [36]. As the trading bots use the sentiment as another variable for prediction, it could weigh it less if classified as less trustworthy. A better approach would be to find the malicious social bots beforehand and apply the sentiment analysis only based on useful social media data.

2.2 Topic Modelling

We conducted a topic modelling analysis to explore research trends in social bots and disinformation literature. Instead of proceeding with a systematic or scoping review [35], identifying the main lines of thought, the methods in use, and the significance of evidence found, we aim to explore the main subjects addressed in related literature collected in a broad topic-wise manner. The strategy is comparable to a rapid review because the collection of evidence presented in the papers is not in focus. Unlike a rapid review done manually by researchers, we argue that a topic modelling approach can be less susceptible to selection bias and more cost-effective since the measures are automated.

According to Blei [1], "topic modelling algorithms are statistical methods that analyse the words of the original texts to discover the themes that run through them, how those themes are connected, and how they change over time". This definition directed us to analyse the topics given the following questions:

1. What are some of the most prominent topics in the literature?
2. How are those topics connected?
3. How likely were those topics addressed over time?

Topic models require no labelling or annotations *a priori*, but instead a list of documents from which we might discover meaningful topics [1,2]. As a first step, we conducted a systematic search in three widely used search engines, namely Web of Science, Scopus, and IEEE. We search within title, abstract, and keywords for papers that include terms, synonyms, or related expressions around bots, finance, social media, and disinformation. The query was built using the terms detailed in Table 1.

Table 1. .

Term	Synonyms and related terms
Bots	(bot$ OR algorithm* OR automat* OR "artificial intelligence")
Finance	(financ* OR trading)
Social media	("social media" OR twitter OR facebook OR youtube OR reddit OR telegram)
Disinformation	(disinformation OR manipulat*)

The query was performed in January 2022. A total of 37 results were found, 7 from Web of Science, 14 from Scopus, and 16 from IEEE. We filtered English-written journal articles or conference proceedings. We did not filter results given a specific year range, and the ultimately selected papers were published between 2016 and 2022. We screened abstracts for relevance, and as a result, 19 papers composed the final corpus.

Topic modelling analyses are often done in extensive corpora, sometimes more than 1000 papers, producing tens or hundreds of topics. Even though our corpus is much smaller, we argue that fewer topics appear prominently with fewer documents, thus being more easily discovered and comprehended. When too many topics are analysed beyond the words that compose them, with aspects such as the connectivity to other topics and how they were addressed over time, it is hard to interpret and keep them in the working memory while reading.

To ensure relevant topic vocabulary, we preprocessed these 19 papers by tokenizing the documents, considering only the predicted words as adjectives, verbs, nouns, and proper nouns. These are more likely to be meaningful words when representing topics. The tokens were considered in their lemmatized forms. We used a stop words list to remove other irrelevant words that remained. After preprocessing, we vectorize the documents using TF-IDF, considering all alphanumeric tokens in lowercase representation and bi-gram term frequencies [43].

We empirically chose to use Non-negative Matrix Factorization (NMF, [27]) for the topic modelling algorithm, using the default implementation provided in the *scikit-learn* package [38]. To choose the number of topics, we empirically inspected the resulting topics, testing integers from 10 to 5 and identifying the number that generates less redundancy across topics by looking at the main words. Starting from 10, we iteratively tried a smaller number if two topics

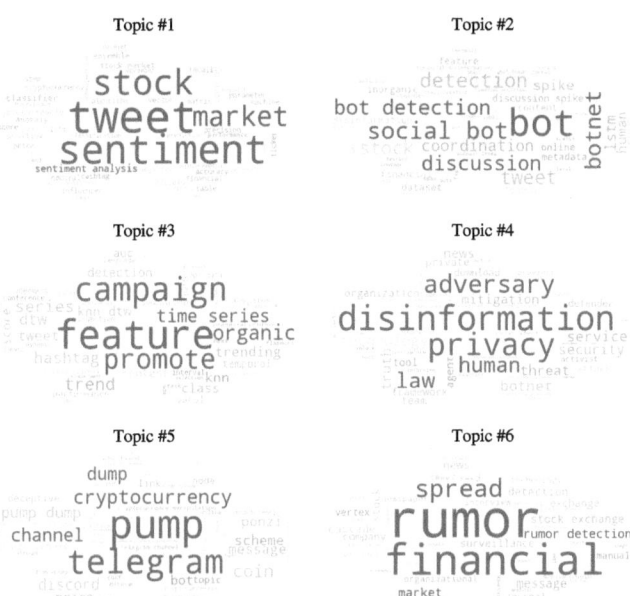

Fig. 1. Topics in word clouds. The top 5 words are highlighted in royal blue. (Color figure online)

or more mentioned one same word (e.g., "bot") in the top 3 most important words. With more than ten topics for nineteen papers, topics would be too fine-grained and less than five too coarse. Following this procedure, the number that generated a less redundant set of topics was $n = 6$.

2.2.1 What are Some of the Most Prominent Topics in the Literature?

In a small dataset such as ours, a small number of words concentrate most of the importance. In Fig. 1, we draw word clouds with the 50 words with the highest importance on each of the six topics. Notice that Topic #1 relates TWITTER to SENTIMENT, STOCK, and MARKET, but also includes some terminology from classification algorithms. Thus, we labelled this one as *"Sentiment analysis"*, discussed mainly by [9, 14, 15, 19, 54]. Topic #2 explicitly focuses on *"Bot detection"*, and relates in secondary importance to terms as INORGANIC, DISCUSSION, TWEET, and COORDINATION. This topic has the leading contributors [22, 24, 26, 44, 47, 49]. Topic #3 might be related to (social media) *"Campaign promotion"*, by relating the words FEATURE, TIME SERIES, ORGANIC, CAMPAIGN, and PROMOTE as of first importance, and of second importance a mixture of classification algorithms and social media related terms (e.g., HASHTAG, CONTENT). This topic was discussed mainly by [13–15, 49, 53].

Topic #4 was labelled as *"Disinformation and privacy"* and relates PRIVACY, LAW, HUMAN, ADVERSARY as one of the focus of discussion in a relatively high

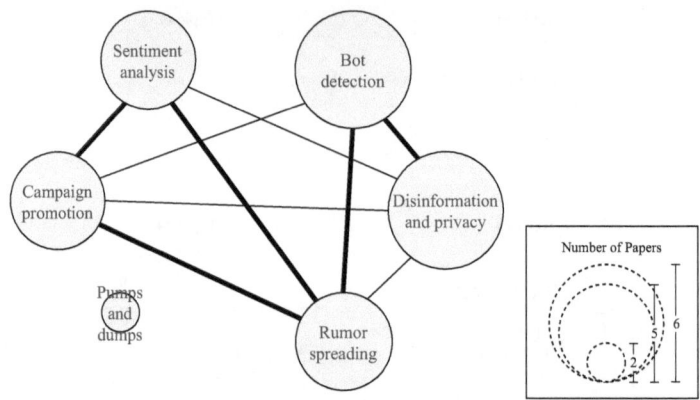

Fig. 2. Connectivity of topics. The size of the nodes representing the number of papers that surpass the threshold level, and the edge's thickness representing the number of papers that discuss both topics.

number of papers [14, 20, 22, 25, 26, 30]. Topic #5 seems to indicate that PUMP is mainly related to INVITE, TELEGRAM, CHANNEL, and CRYPTOCURRENCIES, but also PONZI and SCHEME as relevant terms; we labelled it as *"Pumps and dumps"*, which was discussed by [33, 37]. Finally, Topic #6 describes "Rumour spreading" in more general terms, including STOCK EXCHANGE, MARKET, and DETECTION, but also NEWS, CASCADE as interesting words of secondary importance. This lastly mentioned topic was examined by [14, 31, 45, 49, 54].

2.2.2 How are Those Topics Connected?

For each document, we considered the event of co-occurrence of two or more topics when there are two or more topic values above a threshold of 0.08, representing the 75% quantile level. The connectivity between topics can be seen in Fig. 2.

The biggest nodes are those topics with more papers focused on them. Also in the graph of Fig. 2, the connectivity is more robust when edges are thick (e.g., between *"Rumour spreading"* and *"Sentiment analysis"*, *"Disinformation and privacy"* and *"Bot detection"*). Notice that the topic *"Pumps and dumps"* did not connect to the other topics, probably due to the low number of papers ($n = 2$) included in our literature base that surpasses the threshold level. However, also, this might indicate a research gap where more papers should discuss the interplay between cryptocurrency-related media with campaign promotion, disinformation and privacy, and rumour spreading.

2.2.3 How Likely Were Those Topics Addressed over Time?

To explore the contributions over time by topic, we summed the values assigned to all documents by topic and year. In Fig. 3, one can see that the single document published in 2016 contributed to only the *"Campaign promotion"* topic.

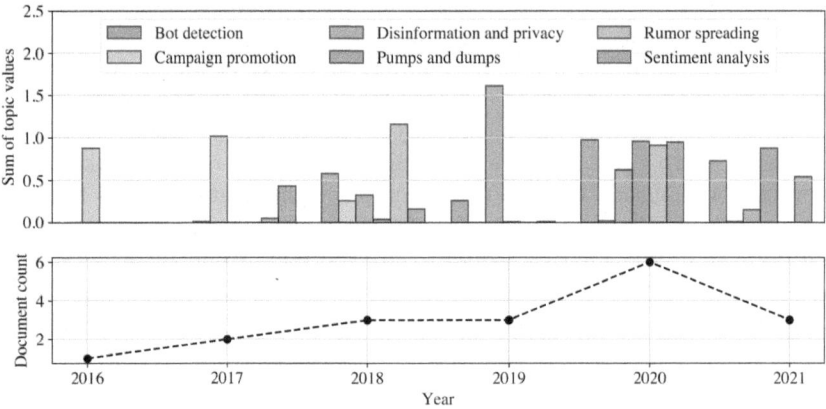

Fig. 3. Contributions to each topic over time. In the top, the sum of topic values from all papers for each topic and in the bottom, the amount of documents by year from 2016 until 2021.

In 2017, the same topic appeared again, but *"Sentiment analysis"* appeared as well. In 2018, *"Rumour spreading"* was the one with the most attention, followed by *"Bot detection"*. Interestingly, in the same year, *"Disinformation and privacy"* also appeared for the first time and was the main focus of attention in 2019. In 2020, there was a peak of publications addressing most of the topics in a balanced manner, but *"Campaign promotion"* seemed to fade out from the discussions a year before. 2020 was also the first time that *"Pumps and dumps"* was prominently addressed by this literature.

2.3 Social Bots in Finance

Since social bots spreading disinformation about finance is a notable topic in our literature review, we will investigate use cases and capabilities in more detail next. Seven papers in our literature base are related to social bots. As can be seen, when considering Fig. 3, all papers were published from 2018 onward. In line with our findings related to Topic #2 in Fig. 1, in all the following studies, social bots related to inorganic campaigns are mainly detected on Twitter by investigating coordination.

Sela et al. [45] analysed how social bots can be used in spreading groups or botnets to promote certain stocks. They simulated the promotion of messages containing information on NASDAQ stocks based on the spread of actual tweets. They find that spreading groups are an excellent method for promoting certain stocks and that social bots are suitable - and much cheaper than commissioning human influencers - for this task. These results were confirmed by Fan et al. [12], who collected tweets about companies listed on the London stock exchange, covering three years. Using simple heuristics to identify social bots (e.g., tweeting at night), they found a significant relationship between bot tweets' sentiment and stock returns, volatility, and trading volume.

This insight is in line with the findings of Cresci et al. [7], who collected tweets mentioning stocks listed on the NASDAQ website in their work. By using the bot detection strategy developed by Cresci et al. [6], they revealed that bots were used to spread tweets containing the names of low market capitalization stocks alongside names of high market capitalization stocks to increase their value artificially. In Cresci et al. [8], they showed that large networks of social bots were used for this strategy, i.e., the bots would retweet these messages synchronously to stimulate increased stock interest in lower market capitalization stocks. Lastly, this research group showed that bots have significantly fewer social ties than humans and thus seem less trustworthy [48]. Therefore, the authors concluded that these bots most likely deceive trading algorithms more quickly than actual human investors [47]. Nevertheless, considering our finding that the Wall Street Journal also observes sentiment on Twitter, it may be possible that on a higher level, also human investors can be influenced by this type of campaign.

As can be seen, when considering Topic #5 in Fig. 1, stocks and cryptocurrencies were the targets of social bot activities. Nizzoli et al. [37] investigated disinformation campaigns on Twitter, Telegram, and Discord by identifying social bots on Twitter via the Botometer developed by Yang et al. [59]. Results indicate that bots infiltrate mostly Twitter and Telegram: 75% of invitations to private channels related to financial information were posted by bots. New investors are lured into private channels where cheaper, unknown cryptocurrencies are offered to increase their value artificially [37]. Similar pump-and-dump strategies were revealed by Mirtaheri et al. [33], who also used the Botometer to show that inferior cryptocurrencies were first promoted by bots on Twitter and then offered via Telegram channels. However, note that the Botometer produces unreliable results, especially when the bots are disguised in a sophisticated way [5]. Since it focuses on analysing single accounts' behaviour, the Botometer misses coordinated actions of more extensive networks or classifies single human users as social bots [41].

Although social bots are used in different areas of the financial industry, the general strategy is similar: bots are being used to spread simple or already created messages to fool potential credulous human or algorithmic investors into buying low-value products so that the operators of the bots can take advantages of the artificially increased prices.

3   Discussion

In the last sections, we investigated the state of finance and disinformation in connection to AI, primarily social bots. We showed that disinformation has become a severe issue for free information exchange on social media. Financial services rely increasingly on technology, AI, and interconnected information flow. However, investors are also increasingly discovering social media as a source of information about what others think about certain stocks; especially since COVID-19. Thus, misinformation and disinformation campaigns like pump-and-dump schemes pose a severe threat to an open and unbiased trading and information exchange environment online.

Even without the technology-aided artificial spread of disinformation on social media, investors cope with many biases, like the continued influence bias and limited attention spans. Therefore, when they are negatively influenced by social bots spreading disinformation, it has serious consequences: they may lose their trust in services, information, and products. The problem is increasingly addressed in research, as seen in our topic modelling analysis: the topic of (social media-related) "sentiment analysis" is closely connected to "campaigns promotion" and "rumour-spreading," indicating that the search for campaigns influencing the attitude of investors towards certain stocks is examined. This threat is also reflected in the close connection of the topics "bot detection, " "misinformation, " and "rumours" in Fig. 2.

Our topic modelling analysis could spot some literature gaps that need further research. For example, we observed that the topic "pumps and dumps" was not frequently associated with other topics, possibly indicating the need for papers to debate the interplay between social media usage and cryptocurrency manipulation, an essential theme for regulatory bodies to address knowledgeably. Moreover, we did not find any paper that discusses both "sentiment analysis" and "bot detection" in more depth. By extracting sentiment-related variables, efforts can be made to understand how bots behave online, e.g., when they post which content or react to a particular stimulus. By widening the focus of bot detection, findings from this behavioural analysis would close this research gap.

Nevertheless, we showed that social bots used momentarily are algorithms spreading predefined messages written by humans. The purpose of these bots is to spread rumours about specific stocks or cryptocurrencies to fool users into investing in them. Thus, a distorted view makes certain financial products seem more attractive than they are. They fool not only human investors but also automated trading agents by evaluating the current sentiment and the popularity of certain stocks on social media. Ultimately, using social bots to spread misinformation and disinformation about investment opportunities raises many sensitive ethical and regulatory questions, which we will discuss in the following.

3.1 Ethical Considerations

A pressing issue is the distinction between sharing volatile financial knowledge on social media platforms and conducting disinformation campaigns with clear motives to fluctuate the market. Beyond all questions, spreading disinformation intentionally and actively as false information propagated to harm others is unethical and should be banned from online communities.

However, what about misinformation and malinformation? For example, if users believe that a company's stock value will plummet and act upon their knowledge, sharing this information with other investors will be morally justifiable. Following the definitions of Wardle [57], if the information is inaccurate, it would be misinformation. Consequently, where is the line between believing in something others may consider untrue and actual misinformation? Further, regarding malinformation: if the information is spread, leading to a stock market crash that harms many investors, this can also be seen as unethical, although the information might be accurate. Hence, it is necessary to specify the terms defined by Wardle [57] in more detail. Disinformation must be fought, independent of whether social bots or human agents propagate it. Nevertheless, punishing the spread of misinformation by users who have no intent to harm others but only express their opinion is an ethical problem that must be examined in more detail in the future. The same is true for malinformation. No one can argue against publishing accurate information, but if it leads to severe consequences, like crashes of stocks or entire markets, it would again lead to an ethical dilemma worth investigating.

Additionally, the role and influence of social bots must be assessed further. On the one hand, they are a facile and easy-to-use tool. They diffuse rumours about specific stocks or cryptocurrencies to fool users into investing in them, thus creating a distorted view. Nevertheless, on the other hand, human users can do the same: influencers send messages to their followers to promote specific products they were paid to sell. Many podcasts or experts spread alleged insider news about investment opportunities in finance. For example, the entrepreneur Elon Musk caused fluctuations in Bitcoin prices by disallowing users to buy a Tesla with Bitcoins. Thus, other users who see him as a role model were tempted to sell their Bitcoins, too[1]. Besides famous internet influencers, groups of investors (or "whales") can cause variations in the market in several other ways by using social media. Kamps et al. [21] suggest that anomaly detection techniques, which can be partly considered an AI tool, can help detect and predict pumps-and-dumps. Therefore, social bots or AI, in general, seem not to be the main problem; instead, humans misbehave in possessing powerful tools for social networks with ineffective or obsolete regulation. Consequently, social media platform moderators must weigh the benefit of deleting disinformation or banishing alleged fake accounts on social media platforms and risk restricting the right to freedom of speech of users.

Some of these problems can eventually be solved with sensible and reasonable regulation. It can be defined which types of practices can be considered spreading misinformation and what framework all stakeholders and parties should follow. For example, when Wardle's [57] definitions regarding dis-, mis- or malinformation are applied to the context of financial markets, platform moderators can decide more easily which accounts can be banned from their social network. However, who should be allowed to pass these regulations, e.g., stock exchange supervisors, financial institutions, or governments? How many regulations are required or can be implemented in a meaningful and assertive way? In the following, we want to investigate these questions by exploring possible answers and assessing their ethical and practical value.

---

[1] https://investorplace.com/hypergrowthinvesting/2021/05/elon-musk-ditches-bitcoin-should-you-do-the-same/.

3.2 Regulatory Considerations

Regulations can be used to set boundaries or guidelines for users, platform operators, or the financial industry to reduce the amount of disinformation spread on social media. However, who should have the right to set and enforce those regulations? Generally, there are four possibilities: government and judicial authorities, institutions from the financial industry, digital platform operators, or the community itself, i.e., the users.

First, states or other official authorities may pass a law to fight disinformation, i.e., force platform operators to act against misuse of their services. An example of that is the draft Artificial Intelligence Act (AI Act). In this draft act, forbidden use cases are listed, oriented on how much risk the AI brings to citizens or societies. For instance, "subliminal manipulation resulting in psychical or psychological harm" is stated as one of AI's criminal use cases, in which social bots' disinformation may fall [4]. The draft AI Act has been criticised for not being sufficiently precisely formulated, so that instead of providing guidance to developers or platform operators, the act might make them insecure. Another law, the Digital Service Act (DSA), obliges large online platforms to have a risk management system in place to fight, among others, manipulation on their platforms.

Regulatory institutions of the financial industry, such as stock exchange supervisors may also issue regulations. Since the early 1900 in the US, institutions like the Commodity Future Trading Commission (CFTC) have tried to find a way to fight rumours and disinformation with various acts and regulations [34]. Another federal agency in the US that is responsible for "protecting investors, maintaining fair, orderly, and efficient markets, and facilitating capital formation" [52] is the US Securities and Exchange Commission (SEC). As mentioned in the discussion, the SEC issued an investor alert about using social media platforms to perform market manipulation schemes. Furthermore, the SEC has prosecuted multiple people with financial crimes relating to market manipulation, market manipulation, false tweets, fake websites, and the dark web in recent years[2]. With the growing threat of utilizing social bots to mislead markets, government agencies such as the SEC may get more involved in combating cyber financial crimes in the future. Nevertheless, fighting rumours is a never-ending task, so frameworks or regulations must be updated to keep track of technology. Finding the correct balance between regulating financial markets and avoiding the spread of rumours and disinformation while protecting the right of investors and markets to operate freely will be a difficult task in the future.

Further, regulations can be enforced by platform operators. Twitter, for example, disallows the use of fake identities and the spread of disinformation in their terms of services [50]. However, the companies must weigh their economic interest against public pressure: on the one hand, the more users they have, the more significant the network and the more money can be made by selling advertisements. Nevertheless, on the other hand, negative news about

---

[2] https://www.sec.gov/spotlight/cybersecurity-enforcement-actions.

disinformation damages the platform's image. Thus, companies may not be the best authority to enforce fair and effective regulations against fake identities and disinformation.

Lastly, in each community, even an online community, the ethical rules of societies must be obeyed. These are made and enforced by other users who either report or argue against fake identities or misbehavior. Crowdsourcing is a technique to detect social bots: groups of users or experts can be asked to investigate datasets from social networks. Overall, they perform excellently, although human investigators are expensive and slow [56]. Additionally, ethical rules depend on societies' culture and thus may vary in a worldwide social network. Thus, other users may also be unsuitable for passing and enforcing regulations.

Overall, different possibilities exist to pass regulations by different authorities. For example, the public pressure on social media forces companies to install disinformation detection mechanisms[3] Similarly, regulations enforced by states or institutions do not seem to have a long-lasting effect, are vaguely formulated, and need to be updated continuously. The question remains whether these not-so-clear regulations can be effective in practice [10].

Moreover, addressing disinformation with regulation brings about the threat of over-regulation, which is disadvantageous for markets and users. We showed that disinformation campaigns impact trading volume but are less significant than legitimate news releases. Furthermore, studies have demonstrated that the markets tend to correct themselves after manipulation attempts. Thus, too many regulations may diminish the investor's trust in the self-correction capabilities of markets.

The fight against social bots and disinformation raises many ethical problems and questions regarding possible regulations. Especially in the financial industry, small actions of malicious users may harm thousands of others. In 2013, for example, a user hacked the news site Fox News and published an article that the at the time president of the US, Barack Obama, died in an accident, causing the stock markets to crash[4]. Thus, regulations or safeguards may still be necessary to protect many investors. However, they must be formulated, enforced, and updated carefully to achieve the right effect.

4   Conclusion

This paper discussed, in the context of financial disinformation, research trends and ethical challenges associated with social bots. We started by providing the context in which social bots became relevant to financial systems and exploring what kind of public debates it consequently raises. By relating to academic literature and events in recent history, we discussed standard practices in which inorganic or fraudulent social media campaigns were used to manipulate market

---

[3] https://www.nytimes.com/2020/06/11/technology/twitter-chinese-misinformation.html.

[4] https://business.time.com/2013/04/24/how-does-one-fake-tweet-cause-a-stock-market-crash/.

sentiment, stocks, and users' opinions with the help of social bot-related technologies. Furthermore, by examining previous research, we can conclude that there is a potential danger of manipulating financial markets as a whole and the effect of disinformation campaigns on especially retail investors. This issue will become more dangerous with the technological advancement of social bots and the availability of state-of-the-art algorithms for fraudsters to manipulate markets.

We leveraged a topic modelling analysis on scientific papers to explore research trends. Our topic models facilitated a bird's-eye view of the most prominent topics in the literature, how they are connected, and how much the literature addressed them over time. The six topics we found addressed sentiment analysis, bot detection, campaign promotion, disinformation and privacy, pumps and dumps, and rumour spreading. By analysing the topics, we could find some interesting patterns. For example, we identified a research gap in the interplay between cryptocurrency-related media and other topics, such as campaign promotion, disinformation and privacy, and rumour spreading. Even though the particular pieces of evidence in the analysed papers were not the focus of our topic modelling, a more profound analysis of the recently published papers can make future research identify other relevant aspects of discussion to inform regulation bodies.

The relationship between social media manipulation and the progress of AI was examined in detail. We suggested that a more accurate perspective of AI and its association with social bots and disinformation is crucial to developing effective regulation. We found that highly accurate AI methods, such as sentiment analysis, deep fakes, anomaly detection, and trading time series prediction might present different risks. On the one hand, AI methods enable the development and increase the efficacy of mistrusted agents, while on the other hand, they can be used to detect suspicious finance-related social media posts and accounts. Ultimately, we reviewed the risks and potentials of regulations to reveal research gaps and opportunities for future works.

References

1. Blei, D.M.: Probabilistic topic models. Commun. ACM **55**(4), 77–84 (2012)
2. Chang, J., Gerrish, S., Wang, C., Boyd-Graber, J.L., Blei, D.M.: Reading tea leaves: how humans interpret topic models. In: Advances in Neural Information Processing Systems, pp. 288–296 (2009)
3. Clarke, J., Chen, H., Du, D., Hu, Y.J.: Fake news, investor attention, and market reaction. Inf. Syst. Res. **32**(1), 35–52 (2021). https://doi.org/10.1287/isre.2019.0910
4. European Commission: Artificial Intelligence Act. Regulation of the European Parliament and of the Council, vol. 0106 (2021)
5. Cresci, S.: A decade of social bot detection. Commun. ACM **63**(10), 72–83 (2020). https://doi.org/10.1145/3409116

6. Cresci, S., Di Pietro, R., Petrocchi, M., Spognardi, A., Tesconi, M.: Social finger-printing: detection of spambot groups through DNA-inspired behavioral modeling. IEEE Trans. Dependable Secure Comput. **15**(4), 561–576 (2017). https://doi.org/10.1109/TDSC.2017.2681672

7. Cresci, S., Lillo, F., Regoli, D., Tardelli, S., Tesconi, M.: $FAKE: evidence of spam and bot activity in stock microblogs on Twitter. In: Proceedings of the Twelfth International AAAI Conference on Web and Social Media, ICWSM 2018, pp. 580–583. Association for the Advancement of Artificial Intelligence, Stanford, CA, USA (2018)

8. Cresci, S., Lillo, F., Regoli, D., Tardelli, S., Tesconi, M.: Cashtag piggybacking: uncovering spam and bot activity in stock microblogs on Twitter. ACM Trans. Web **13**(2), 1–27 (2019). https://doi.org/10.1145/3313184

9. Doğan, M., Metin, Ö., Tek, E., Yumuşak, S., Öztoprak, K.: Speculator and influencer evaluation in stock market by using social media. In: 2020 IEEE International Conference on Big Data (Big Data), pp. 4559–4566. IEEE (2020)

10. Ebers, M., Hoch, V.R., Rosenkranz, F., Ruschemeier, H., Steinrötter, B.: The European commission's proposal for an artificial intelligence act-a critical assessment by members of the robotics and AI law society (rails). J **4**(4), 589–603 (2021)

11. Fagni, T., Falchi, F., Gambini, M., Martella, A., Tesconi, M.: TweepFake: about detecting deepfake tweets. PLOS ONE **16**(5) (2021). https://doi.org/10.1371/journal.pone.0251415

12. Fan, R., Talavera, O., Tran, V.: Social media bots and stock markets. Eur. Financ. Manag. **26**(3), 753–777 (2020). https://doi.org/10.1111/eufm.12245

13. Ferrara, E., Varol, O., Menczer, F., Flammini, A.: Detection of promoted social media campaigns. In: Proceedings of the Tenth International AAAI Conference on Web and Social Media. ICWSM 2016, vol. 10. AAAI, Cologne, Germany (2016)

14. Geçkil, A., Müngen, A.A., Gündogan, E., Kaya, M.: A clickbait detection method on news sites. In: 2018 IEEE/ACM International Conference on Advances in Social Networks Analysis and Mining (ASONAM), pp. 932–937. IEEE (2018)

15. Golmohammadi, K., Zaiane, O.R.: Sentiment analysis on twitter to improve time series contextual anomaly detection for detecting stock market manipulation. In: Bellatreche, L., Chakravarthy, S. (eds.) DaWaK 2017. LNCS, vol. 10440, pp. 327–342. Springer, Cham (2017). https://doi.org/10.1007/978-3-319-64283-3_24

16. Grant, S.M., Hodge, F.D., Seto, S.C.: Can prompting investors to be in a deliberative mindset reduce their reliance on fake news? SSRN (2021)

17. Grimme, C., Preuss, M., Adam, L., Trautmann, H.: Social bots: human-like by means of human control? Big Data **5**(4), 279–293 (2017). https://doi.org/10.1089/big.2017.0044

18. Hajli, N., Saeed, U., Tajvidi, M., Shirazi, F.: Social bots and the spread of disinformation in social media: the challenges of artificial intelligence. Br. J. Manag. 1–16 (2021)

19. Ibrahim, A.: Forecasting the early market movement in bitcoin using twitter's sentiment analysis: an ensemble-based prediction model. In: 2021 IEEE International IOT, Electronics and Mechatronics Conference (IEMTRONICS), pp. 1–5. IEEE (2021)

20. Isle, B., Smith, T.: Real world examples suggest a path to automated mitigation of disinformation. In: 2018 IEEE International Conference on Big Data (Big Data), pp. 4408–4412. IEEE (2018)

21. Kamps, J., Kleinberg, B.: To the moon: defining and detecting cryptocurrency pump-and-dumps. Crime Sci. **7**(1), 1–18 (2018)
22. Khaund, T., Kirdemir, B., Agarwal, N., Liu, H., Morstatter, F.: Social bots and their coordination during online campaigns: a survey. IEEE Trans. Comput. Soc. Syst. 1–16 (2021). https://doi.org/10.1109/TCSS.2021.3103515. https://ieeexplore.ieee.org/document/9518390/
23. Kogan, S., Moskowitz, T.J., Niessner, M.: Fake news: evidence from financial markets. SSRN Electron. J. (2018). https://doi.org/10.2139/ssrn.3237763
24. Kudugunta, S., Ferrara, E.: Deep neural networks for bot detection. Inf. Sci. **467**, 312–322 (2018)
25. Land, M.K., Aronson, J.D.: Human rights and technology: new challenges for justice and accountability. Ann. Rev. Law Soc. Sci. **16**, 223–240 (2020)
26. Lange, T., Kettani, H.: On security threats of botnets to cyber systems. In: 2019 6th International Conference on Signal Processing and Integrated Networks (SPIN), Noida, India, pp. 176–183. IEEE (2019). https://doi.org/10.1109/SPIN.2019.8711780. https://ieeexplore.ieee.org/document/8711780/
27. Lee, D.D., Seung, H.S.: Learning the parts of objects by non-negative matrix factorization. Nature **401**(6755), 788–791 (1999)
28. Lehrer, S., Xie, T., Zhang, X.: Social media sentiment, model uncertainty, and volatility forecasting. Econ. Model. **102**, 105556 (2021) https://doi.org/10.1016/j.econmod.2021.105556. https://www.sciencedirect.com/science/article/pii/S0264999321001450
29. Lin, T.C.W.: The new market manipulation. Emory Law J. **66**, 1253–1314 (2017)
30. Mahmood, S.: The anti-data-mining (ADM) framework-better privacy on online social networks and beyond. In: 2019 IEEE International Conference on Big Data (Big Data), pp. 5780–5788. IEEE (2019)
31. Majumdar, A., Bose, I.: Detection of financial rumors using big data analytics: the case of the Bombay stock exchange. J. Organ. Comput. Electron. Commer. **28**(2), 79–97 (2018)
32. Mao, Y., Wei, W., Wang, B., Liu, B.: Correlating S&P 500 stocks with twitter data. In: Proceedings of the First ACM International Workshop on Hot Topics on Interdisciplinary Social Networks Research, pp. 69–72. Association for Computing Machinery, New York (2012). https://doi.org/10.1145/2392622.2392634
33. Mirtaheri, M., Abu-El-Haija, S., Morstatter, F., Steeg, G.V., Galstyan, A.: Identifying and analyzing cryptocurrency manipulations in social media. IEEE Trans. Comput. Soc. Syst. **8**(3), 607–617 (2021). https://doi.org/10.1109/TCSS.2021.3059286. https://ieeexplore.ieee.org/document/9371307/
34. Mixter, C., Flores, L., Lewis, M.: The War on Rumors at the SEC and CFTC. Securities Regul. Law Rep. **40**(42), 5 (2008)
35. Munn, Z., Peters, M.D., Stern, C., Tufanaru, C., McArthur, A., Aromataris, E.: Systematic review or scoping review? Guidance for authors when choosing between a systematic or scoping review approach. BMC Med. Res. Methodol. **18**(1), 1–7 (2018)
36. Nan, A., Perumal, A., R., .Z.O.: Sentiment and knowledge based algorithmic trading with deep reinforcement learning. CoRR abs/2001.09403 (2020)
37. Nizzoli, L., Tardelli, S., Avvenuti, M., Cresci, S., Tesconi, M., Ferrara, E.: Charting the landscape of online cryptocurrency manipulation. IEEE Access **8**, 113230–113245 (2020). https://doi.org/10.1109/ACCESS.2020.3003370

38. Pedregosa, F., et al.: Scikit-learn: machine learning in python. J. Mach. Learn. Res. **12**, 2825–2830 (2011)

39. Petratos, P.N.: Misinformation, disinformation, and fake news: cyber risks to business. Bus. Horiz. **64**(6), 763–774 (2021). https://doi.org/10.1016/j.bushor.2021.07.012

40. Petratos, P.N.: Misinformation, disinformation, and fake news: cyber risks to business. Bus. Horiz. **64**(6), 763–774 (2021). https://doi.org/10.1016/j.bushor.2021.07.012. https://www.sciencedirect.com/science/article/pii/S000768132100135X

41. Rauchfleisch, A., Kaiser, J.: The false positive problem of automatic bot detection in social science research. PLoS ONE **15**(10), e0241045 (2020). https://doi.org/10.1371/journal.pone.0241045

42. Renault, T.: Market manipulation and suspicious stock recommendations on social media. SSRN Electron. J. (2017). https://doi.org/10.2139/ssrn.3010850

43. Sammut, C., Webb, G.I.: TF–IDF. In: Sammut, C., Webb, G.I. (eds.) Encyclopedia of Machine Learning, pp. 986–987. Springer, Boston (2010). https://doi.org/10.1007/978-0-387-30164-8_832

44. Sela, A., Cohen-Milo, O., Kagan, E., Zwilling, M., Ben-Gal, I.: Using connected accounts to enhance information spread in social networks. In: Cherifi, H., Gaito, S., Mendes, J.F., Moro, E., Rocha, L.M. (eds.) COMPLEX NETWORKS 2019. SCI, vol. 881, pp. 459–468. Springer, Cham (2020). https://doi.org/10.1007/978-3-030-36687-2_38

45. Sela, A., Milo, O., Kagan, E., Ben-Gal, I.: Improving information spread by spreading groups. Online Inf. Rev. **44**(1), 24–42 (2019). https://doi.org/10.1108/OIR-08-2018-0245

46. Suwajanakorn, S., Seitz, S.M., Kemelmacher-Shlizerman, I.: Synthesizing Obama: learning lip sync from audio. ACM Trans. Graph. (ToG) **36**(4), 1–13 (2017)

47. Tardelli, S., Avvenuti, M., Tesconi, M., Cresci, S.: Characterizing social bots spreading financial disinformation. In: Meiselwitz, G. (ed.) HCII 2020. LNCS, vol. 12194, pp. 376–392. Springer, Cham (2020). https://doi.org/10.1007/978-3-030-49570-1_26

48. Tardelli, S., Avvenuti, M., Tesconi, M., Cresci, S.: Detecting inorganic financial campaigns on twitter. Inf. Syst. 101769 (2021)

49. Tardelli, S., Avvenuti, M., Tesconi, M., Cresci, S.: Detecting inorganic financial campaigns on Twitter. Inf. Syst. **103**, 101769 (2022). https://doi.org/10.1016/j.is.2021.101769

50. Twitter Help Center: Report impersonation accounts (2022). https://help.twitter.com/en/safety-and-security/report-twitter-impersonation

51. US Securities and Exchange Commission: Investor alerts and bulletins (2015). https://www.sec.gov/oiea/investor-alerts-bulletins/ia_rumors.html

52. US Securities and Exchange Commission: What we do (2021). https://www.sec.gov/about/what-we-do

53. Varol, O., Ferrara, E., Menczer, F., Flammini, A.: Early detection of promoted campaigns on social media. EPJ Data Sci. **6**(1), 1–29 (2017). https://doi.org/10.1140/epjds/s13688-017-0111-y

54. Vilas, A.F., Redondo, R.P.D., García, A.L.: The irruption of cryptocurrencies into twitter cashtags: a classifying solution. IEEE Access **8**, 32698–32713 (2020)

55. Vosoughi, S., Roy, D., Aral, S.: The spread of true and false news online. Science **359**(6380), 1146–1151 (2018). https://doi.org/10.1126/science.aap9559

56. Wang, G., et al.: Social Turing Tests: Crowdsourcing Sybil Detection. arXiv (2013)

57. Wardle, C.: The need for smarter definitions and practical, timely empirical research on information disorder. Digit. Journal. **6**(8), 951–963 (2018)
58. Wojtowicz, Z., Chater, N., Loewenstein, G.: Boredom and flow: an opportunity cost theory of attention-directing motivational states. SSRN (2021)
59. Yang, K.C., Varol, O., Hui, P.M., Menczer, F.: Scalable and generalizable social bot detection through data selection. In: Proceedings of the AAAI Conference on Artificial Intelligence, vol. 34, no. 01, pp. 1096–1103 (2020). https://doi.org/10.1609/aaai.v34i01.5460

Data Protection and Surveillance: Novel Pathways of an Ethical Data Economy

Syeda Amna Sohail[1], Michaël Grauwde[2], and Julian von Lilienfeld-Toal[3(✉)]

[1] Department of Data Management and Biometrics (DMB), Faculty of Electrical Engineering, Mathematics and Computer Science (EEMCS), University of Twente, Enschede, The Netherlands
`s.a.sohail@utwente.nl`
[2] Leiden Institute of Advanced Computer Science, Leiden University, Leiden, The Netherlands
`m.n.j.grauwde@umail.leidenuniv.nl`
[3] Department of Information Systems, Westfälische Wilhelms-Universität Münster, Münster, Germany
`julian.lilienfeld@gmail.com`

Abstract. Today, the data ecosystem is so technically advanced and embedded in our daily lives that everything we do generates value for multiple stakeholders in the data economy. This overlapping of technology, economy, and society makes it incumbent upon national and pan-European authorities to safeguard often conflicting interests of all stakeholders in the best interest of all. However, unfortunately, the laws and regulations may be imposed, but are often not followed in their genuine sense. The underlying reasons include, amongst others, the absence of apt (technical and organizational) solutions and knowledgeable, well-equipped oversight bodies. This paper examines the implication of FACT principles, FAIR principles, and Open Personal Data Stores for (potential) novel pathways for an ethical data economy. For clarity, the research work comprises three dimensions, namely ethical (from the individuals' perspective), economic (from the enterprises' perspective), and legal (focusing on the informed data collection perspective). For evaluation, the literature review and content analysis are performed in each dimension, and recommendations are presented for efficient and effective ethics inclusion in today's data economy. The research will facilitate the concerned authorities and key stakeholders in implementing recommendations for an improved ethical data economy.

1 Introduction

The data economy comprises producing, sharing, and storing data to extract value [38]. Data as a valuable commodity interconnects multiple information systems in the form of a data economy ecosystem [26]. Data as fuel for the data economy facilitates private/public bodies across domains in ensuring better products and improved services [37]. However, with the rapid advancement of data analytics' tools and techniques, public and private bodies (organizations/individuals)

J. Bayer and C. Grimme (Eds.): AI, Human Rights and Ethics, LNAI 14400, pp. 96–112, 2025.
https://doi.org/10.1007/978-3-031-52082-2_7

lagged behind in: ensuring ethical data handling and protecting fundamental human rights [37]. Ethical data handling involves adhering to (legal) set standards [12] to ensure the well-being of the society at large. Here, data handling involves assimilation, structuring, sharing, curation, and reusing data. The research work focuses on ethical data handling, especially concerning AI (artificial intelligence) [1] and ML (machine learning) [33]. The big data cycle involves three phases, namely infrastructure, analysis, and effects [4]. Infrastructure is about collecting, storing and processing data. Analysis is to extract value with useful insights and decision making. Effects are about the effects of those decisions upon the society at large. At each phase, the interests of enterprises (might) clash with that of the individuals and can raise ethical issues. We will touch upon all three phases for an overview and evaluation of ethical measures and in locating novel pathways for an ethical data economy.

Our research question is: what are the novel pathways for an ethical data economy? Our *research objective* is to explore the ethical, financial, and legal dimensions of the data economy; to evaluate the (bias-related) solution measures in the above-mentioned dimensions in ethically handling data; and to verify the legal findings using a case study of cookie banners. Our *research contribution* is identifying ethical, financial, and legal dimensions of the data economy and evaluating the (bias-related) respective solution measures in ethically handling data. The research work also verifies the legal findings using a use case of cookie banners on websites. The following section includes the approach and methodology of the research work. Section 3 presents the literature review in three dimensions comprising ethical, economic and (informed data collection) legal dimensions. Section 4 findings and discussions comprise evaluation of all three dimensions including respective recommendations. Section 5 includes a conclusion at the end.

2   Approach and Methodology

The co-authors collectively formulated the research question based on the research objective via virtual meeting. The three dimensions are ethical, economic, and legal dimensions of data economy. Each author reviewed relevant literature, evaluated the problem area, and recommended solution measures in respective dimensions. For example, the ethical dimension focuses on protecting the fundamental human rights of individuals as data subjects. The recommendations include the FACT (fair, accurate, confidential, and transparent) principles' evaluation with risk assessment of opacity, scale, and damage to the human race from a Responsible Data Science (RDS) perspective. The economic dimension highlights the binary fiscal management approach of the enterprises in an extension of enterprises' respective business goals. The legal dimension excludes minor legal details but focuses on the faulty implementation of the legal requirements such as informed data collection and verifies the problem area with a use case of cookie banners'. For detail see Fig. 1.

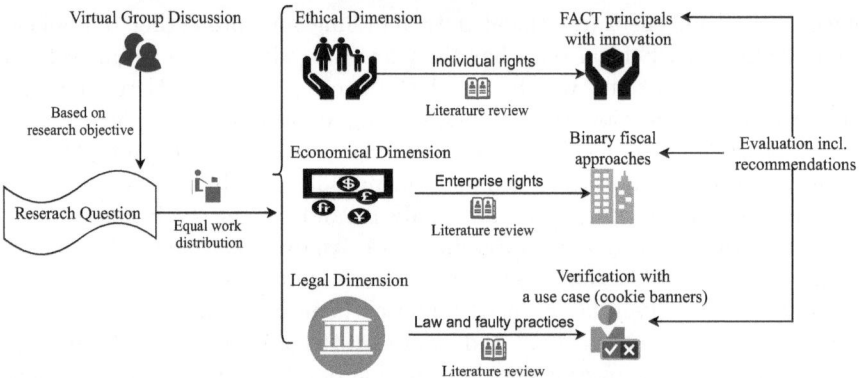

Fig. 1. Research Approach with Methodology

3   Literature Review

The literature review is divided into three subsections, namely ethical data handling, economic data handling and informed data collection as given below:

3.1   Ethical Data Handling

Ethics emerge from fundamental ethical theories such as Kantianism, Utilitarianism, Social contract theory, and Virtue theory [25]. Ethics emphasize our role as responsible community members and urge us to be cautious of our actions and their consequences upon other community members.

Ethics are intrinsically humane yet somewhat beyond intrinsic human behaviour. Therefore, the state legally protects fundamental human rights to regulate human behaviour on the one hand and to limit state powers on the other [37]. Machine learning (ML) and artificial intelligence (AI) oriented big data prejudices have caused irrevocable large-scale harm to humanity, for example, in predictive decision making and selection across domains [33,41]. For a sustainable, ethical data economy, ethics should be legalized (and adhered to) to safeguard the multilevel stakeholders' rights. This includes fundamental human rights' protection from other humans/organizations, small and medium enterprises (SMEs) protection from corporate conglomerates and potent rivals, and underdeveloped and poor nation-states' rights from economically and technically advanced countries.

The literature review reveals that the research studies are divisible into ethics concerning enterprises and ethics relating to individuals' fundamental rights. Ethically, the enterprises clearly focus on transparency for accountability's sake, accuracy for data utility sake, and customer care to serve service-dominant logic [13,43,53]. Ethical concerns from individuals' perspective include confidentiality, privacy, informed consent intake, personal autonomy (agency and choice), and fair (equal) treatment [43,47].

3.2   Economic Data Handling

Current Situation
In the past two years, organizations have recovered and re-strategised due to the COVID-19 pandemic. Transition to an AI-focused strategy was seen as an opportunity, with many still wondering if they can take advantage of the data economy [28]. Data has revolutionized humanity and the role of organizations and companies engaging in data collection. Google Flu Trends, a service that tracks influenza outbreaks, is one example of the benefits of data handling [49]. However, data has also become the lifeblood of many businesses such as Facebook and Snapchat that built their business models around data collection while users remain in the dark about their data handling [17, 23]. The tradeoff between what users know about their data and companies' data use is a delicate issue that will be examined in the coming years as companies and governments find a balance between the benefits and costs of collecting data from an economic perspective.

Benefits of Big Data
Studies have found productivity gains, improvements in competitive advantage and efficiency as some benefits of Big Data [8]. Economic benefits lay in risk prediction for industries such as the insurance industry and better prediction of consumer behaviour for retail stores [8]. Furthermore, big data allows governments and scientists to make improvements in domains like census data collection and epidemic diffusion. There are three categories of data that companies refer to: first party (data gathered by a business from their customers), second party (data gathered from a collaboration with other companies), and third party (open data collected by others, like the government) and each type of organization is involved in different data types [8].

Big Data Confusion Individuals that participate in these companies' business practices are often unaware of the data types as well as what's happening with the data. Often, explanations for data handling are bogged down in cookie banners or privacy policies. Often, companies such as Amazon collect both first party and third-party data to know their consumers and the market better; however, their consumers are left in the dark [8]. Big data has also allowed for new industries such as algorithmic trading that analyses a large number of market data and finds opportunities to gather value for companies. These industries have allowed for the rise of many start-ups to rise and growth exponentially [24]. For companies such as Uber and Airbnb, data is vital for them in their ability to give their consumers real-time information to make real-time decisions [8].

The harvesting and analysing of large data sets leads to privacy concerns. Companies have used individuals' information, shared it, and this sharing has led to exposure. And while there are methods of de-identification being used to contain privacy issues, scientists have shown that the data can be re-identified [49]. A problem surrounding data and data sets remains the lack of examination around why there is so much data handling and gathering. While companies and organizations often talk up the advantages that come with large data sets, they often forget to state the costs and concerns with handling such data. As Chiou and Tucker (2017) found, larger data sets often did not improve the quality of searches that search engines provided to their users [11].

Even so, companies argue that collecting data allows to them to improve services and better analyse customer behaviour. Data allows companies and individuals with data access to hold an immense advantage over their competitors. This also applies for vast information over the behaviour and outcomes of various users of these services while users remain in the dark with regard to data collection [5]. Users themselves often fill out informed consent forms without even knowing what they are about and the negative impacts that their choices may have on their lives [5]. Companies often state that they use privacy policies to be transparent as evidence that they are self-regulating, however, users are often unaware what these terms mean. In fact, many website users believe that privacy policies allow their data to remain private [17]. Furthermore, scholars also argue that transparency campaigns such as Facebook allowing users to download all their collected data may be transparent, but users themselves are incapable of taking action in order to change the power balance between themselves and the company even if they are "transparent" [17]. Users continue to find it confusing to engage with privacy policies while companies state that they have done their job in the name of transparency [17]. This has led to what Draper and Turow (2019) name the

"digital resignation" [17].

Incoming Regulation

Slowly, but surely, the concerns around privacy and calls for governments to address concerns have grown. In anticipation of looming regulation and public scrutiny, Apple made changes regarding their iOS (operating software) [48]. Snap Inc and Facebook are two companies that in recent years have made data one of the most important parts of their business model, so may be hurt by these regulations. Apple's privacy policy change, allowing users to opt out of certain ad tracking caused YouTube, Twitter, Facebook, and Snap Inc to lose around 12% of ad revenue in quarters three and four of 2021 [23]. Next to companies like Apple that have started self-regulation, companies will face new regulation that may make business more confusing for them but may even out the power balance somewhat for users.

In the wake of the big data divide, regulation discussions have come to the fore. Around the end of the 2000s, as the EU was discussing a new legal framework for personal data protection, one of the most ardently argued for was the right to be forgotten. This is the right for persons to

"have their data fully removed when it is no longer needed for the purposes for which it was collected"

[6]. While this issue was broadly positively received in the EU, most US commentators were hostile towards this new proposal and had concerns over freedom of expression around privacy [6]. The EU has led the line in regard to data privacy

with the right to erase data supported by the EU Court of Justice in 2014, setting a precedent [35]. The precedent was later included in the GDPR (General Data Protection Regulation) which has led the line for data privacy around the world. And the EU seems to be moving forward with further regulations such as the ePR (ePrivacy Directive) revolving around cookies and cookie compliance to allow users more control over cookie collected data [50]. These new regulations provide a new landscape for users and companies and present an obstacle for the advertising industry. Goldfarb and Tucker [21] found that online banner advertisements would lose their effectiveness from new privacy laws. Privacy laws would cause the reduction of banner advertisements by 65%, which would lead to these companies having to invest far more in advertisements than they currently do [21].

Shifting Realities
These are the economic implications that companies face under new regulations. These regulations will not only impact companies, but a wide range of workers as well. While many users did not sign up for their data to be used in the manner that it currently is, many companies have made data a vital aspect of their business model. This issue now presents a rocky future for businesses in these industries.

In the following, the data collecting entity will be called controller, whereas the entity, whose data is collected, will be called data subject (following the nomenclature of the General Data Protection Regulation (GDPR) [12]).

3.3 Informed Data Collection

One problem regarding data collection can be, that the data subject is not aware or not fully aware of the collection. We will have a look at informed consent as described in the GDPR [12,20,52], as well as the informed consent in practice [32,42,52].

Legal Basis
Because of the unequal nature and the unequal power distribution between data subjects and controllers, as well as their different interests, issues regarding ethics are prone to arise [36]. The unequal power distribution leads to the necessity of a higher power being involved, e.g. a government. The GDPR is the approach of the European Union to ensure

> [...] the protection of natural persons with regard to the processing of personal data [...]

(GDPR, [12]). This means, for example, that a data subject must always give consent to the controller for processing its data. Additionally, this consent to processing is purpose bound (Article 6a, GDPR, [12]). This consent to data collection is one important principle of the GDPR, mainly described in Article 6 and 7. The consent must be freely given, informed and revoke-able at any time [12,20]. Furthermore, it is stated that:

It shall be as easy to withdraw as to give consent.

(GDPR, Article 7 [12]). This was also recently again enforced in France, when Google (Alphabet) and Facebook had to pay fines due to making it hard to opt out of data tracking [10]. These rules also apply to the use of cookies, at which we will have a closer look in Sect. 4.3.

GDPR and Legal Consent in Practice
The GDPR is user-centric, meaning it focuses on the impact on the user, thus providing a more dynamic approach than previous regulations [20]. This leads to very different implementations by controllers [52]. This can be seen, for example, when looking at cookie consent. Whilst some sites offer the user to select and deselect specific cookies (often divided into categories like necessary, marketing, and more), other sites only offer to accept all cookies or to leave the site (see also [32,52]). Those different options are often confusing to users, leading to users just to

click any interaction element that causes the notice to go away

(Utz et al. in [52]) without them thinking about it. Not only may the options be confusing, the controller may make the understanding even harder by using dark patterns to guide users to just accepting [32,42]. This makes the cookie banners basically useless, since their purpose was to make sure the subject would make an informed decision about the collection of his personal data, but the user may just ignore them without giving a thought about the data collection. According to Degeling et al., these cookie banners may not only not help, but may actually cause harm, since the user may feel more private and secure than he actually is [16]. Furthermore, the inflationary use of cookie banners may lead to a certain fatigue in users, mitigating the actual relevance of cookie banners [9]. Many Websites, who use such banners, may comply with the GDPR, but the achievement of the actual goal of the GDPR, informed consent, has to be questioned. Furthermore, as found by Mehrnezhad, websites often start collecting data before the consent is given, a non-compliant practice commonly found [29]. The fatigue induced in users may lead to users *just accepting everything* without actually thinking about it [9,16,52]. Therefore, the effectiveness of the GDPR has to be questioned, especially considering that enforcement of the GDPR is also not trivial [22], and in many cases not happening [29,32]. Therefore, in the case of the cookie banner, the GDPR is not only hard to enforce, but even if it is enforced, the actual effect it has, which for many users is just to accept everything, is not the intended effect, which would have been informed consent.

4   Findings and Discussion: Evaluation of Solution Measures Towards Achieving Novel Pathways for An Ethical Data Economy

4.1   From a Perspective of Individuals as Data Subjects

Today's data economy allows enterprises to assimilate, share, and store big data from augmented and automatic digital means [19]. Big data evaluation facilitates enterprises in gathering valuable insights regarding cost and time-efficient KPIs achievement and sustainable measures to maintain customers' trust. However, big data evaluation poses unethical implications for data subjects (i.e., individuals). Unethical implications include issues relating to data protection, fairness, transparency, confidentiality, integrity, and autonomy [19].

The research work aims to locate the best possible measures to ascertain ethical data handling where no fundamental human (information) rights are violated. For optimal utilization of data pooling across domains by industries and governments alike, so far, the FAIR (findable, accessible, interoperable, and reusable) principles are globally acclaimed and adopted (go-fair.org, bbmri.nl, zonmw.nl). FAIRification requires "data as open as possible and as close as necessary". FAIRification is a process of making metadata FAIR within and across domains [43]. Here, the definition of 'necessary' remains unclear and rests entirely upon the motivation of (public/private) data handling bodies. On the contrary, the ethical handling of big data has still taken a slower pace nationally and globally. The underlying reasons behind rapid FAIRification efforts lie in the facilitation of the modern data economy and overall national/global financial growth [44,45].

Plausible, best-suited, novel pathways for an ethical data economy
In practice, achieving an ethical data economy requires concerted efforts from multiple (private/public) key stakeholders across domains. Key stakeholders include state bodies, enterprises, and academia alike. Ethics oblige the key stakeholders to be responsible towards data subjects in protecting the latter's personally identifiable information (PII) while handling big data. In this wake, a few Dutch academic institutions have gathered and integrated their efforts for ethical data handling under the tag of the Responsible Data Science (RDS) consortium [3]. The Dutch team of the RDS consortium has taken crucial footsteps by identifying and implementing FACT (fair, accurate, confidential, and transparent) principles with new suitable tools and techniques [3,4]. RDS consortium with competent (infield) researchers, data scientists, and experts has identified four crucial factors that are integral for responsible data handling and, in turn, for an ethical data economy that simultaneously ascertain data utility (i.e., precision of data analytics) [43].

The evaluation of the vital FACT principles from an RDS perspective in pursuit of an ethical data economy is given below

Fairness
Fair means prejudice/bias-free data handling where all data subjects are treated equally. The fair in FACT principals should not be confused with the FAIR of FAIRification mentioned above. In essence, fair of FACT principles imply 'impartial' treatment. Equality and justice cease to exist without fair/impartial treatment of consumers. Fairness is the kernel of building a trustworthy relationship with consumers and is legally ordained (EC regulation for AI, 2021) [2]. Here, EC, European Commission, aims to strike a fair balance between socio-economic aspects of AI in the contemporary world of highly interconnected systems networks. Recent stress upon consumers' trust-building also highlights the market's shift from previously adopted goods' output logic to contemporary service-

dominant logic [46]. This logic keeps consumers at a priority as it seeks to integrate consumers' satisfaction/feedback at each phase of decision-making. Service dominant logic prioritizes consumers' trust-building in today's system networks for maintaining (products/services) utility and value co-creation between consumers and enterprises alike [53]. Fairness inundates equality and justice and is endorsed by OECD countries, 2019 with OECD Council Recommendation on Artificial Intelligence and later by G20 [34]. Hence, fairness and equality are considered as vital constituents in the ethical data economy.

Accuracy

Accuracy requires factual data that is free from guesswork and is accurate in its findings. Accuracy is sometimes feared to be biased and should be traded against fairness [33]. For example, statistical algorithms that predict recidivism are trained with historical training data. In this wake, the criminals with recurring offences are categorized as per their race, gender, age, educational and financial backgrounds. The training data feeds similar information to predictive algorithms that are applied to testing data for factor scoring and identification of future (potential) re-offenders. In this situation, the training data is accurate but biased against future offenders of a certain race, gender, age, and financial and educational backgrounds, leading to discriminatory decision-making. Accuracy also entails the validity of the findings considering all possible (internal and external) threats to validity, such as changes in context, population, surroundings, cause, and effects. For example, in data handling, one training data of different populations with different socio-economical backgrounds does not apply to a starkly different population with entirely different backgrounds. Similarly, if training data lacks complete information and guesswork is added to fill the data, it also hampers the accuracy of the data.

Confidentiality

The confidential data requires efficient and effective problem solving while maintaining the secrets of data subjects (i.e., individuals). Privacy includes the confidentiality measures where data subjects must exercise direct/indirect control over what to share, with whom, and to what extent [44]. Confidentiality is an essential metric of Privacy by Policy (PbP) indicators that is decreed upon all enterprises, across domains, to ensure data subjects' privacy. PbP measures comprise availability (of information access), integrity (of data subjects), confidentiality (of data subjects), and accountability of the concerned authorities [46]. ISO accredits, Privacy by Policy of enterprises across domains, as an international standard organization. These measures are further amalgamated with the privacy by design indicators by each organization that seeks ISO accreditation [46].

Transparency

Transparency is to ascertain the authenticity of the data pipeline by making it accessible, interpret-able, and explainable [34].

In today's highly integrated data ecosystem, data handling involves the "authority of the inscrutable" and causes "invisible harms" to data subjects

[33]. Consequently, the data subjects remain wholly uninformed and ignorant of what is being done, by whom, and for what purpose. Right from the moment of data collection, informed consent ensures transparency and elucidates the process it will go through, from data extraction to its saving and reuse (see also Sect. 3.3 and Sect. 4.3). Transparency ensures limited privacy-related issues, adds clarity, and resolves disputes concerning the authenticity of the data pipeline. Moreover, transparency facilitates the responsibility and accountability of concerned authorities. Transparency also inculcates trust and reliability, through interpret-ability and explain-ability, on an individual's part.

The aforementioned FACT principles give an ideal ground for the rest of the member states and other nation-states to follow. Even in the Netherlands, only a handful of academic institutions have united in amalgamating FACT measures for ethical data handling. The process still lags in launching a full-fledged approach where all stakeholders adopt FACT principles towards an ethical data economy. Their authenticity can be further solidified by checking them at the backdrop of Cathy o Neil's opacity, scale, and damage factors from the data subjects' perspective [33]. Overall, the endeavour is slow-paced and requires (oversight) regulatory bodies to keep the key stakeholders on the right track where ethics are followed, both by design and by policy, towards achieving an efficient and effective data economy. Oversights bodies also need to be well-equipped to assess the technical and administrative measures as per ethical requirements for an efficient and effective process evaluation and guidance.

4.2   Enterprise Perspective

Enterprises often claim that they collect data because it is beneficial to the users, and it allows their businesses to create better products. However, as we saw earlier from Chiou and Tucker [11], this claim is not entirely true, rather, they found that for search engines, holding on to data for a longer time, does not make their services any better [11]. Furthermore, organizations often say that if they did not collect data, there would be costs and that users accept the trade-off to be kept in the dark. However, users are also hesitant to hand over their data to companies they do not see as trustworthy or transparent. While users are willing to hand over their data, if they know and trust what they may get in return [30]. Different companies go through different challenges concerning their transparency that they give their users and the trust that their users have in them in return. Companies like Facebook are on the lower side of trustworthiness and this can allow companies which have facilitated more trust, such as Amazon, to gain a competitive advantage regarding their users [30].

New Ideas
In this section we will explore various ways that enterprises can participate in the data economy while also providing a sense of power to the users whose data is being used.

F.A.I.R. Principles: One aspect of data handling that we will examine are the FAIR principles. FAIR is an abbreviation for Findable, Accessible, Interoperable, and Reusable. The FAIR principles can allow scientists and researchers the ability to analyse data about impending pandemics as well as easily share data with fellow collaborators. During the COVID-19 pandemic, FAIR principles have played a major part in the sharing of data. And the National Institute of Health (NIH) advocates for FAIR principles in its ability to share its data on genomic material with international researchers [39]. Data sharing is one of the most important manners for science to progress and allows scientists to communicate and cooperate globally. So adopting open data and FAIR principles should be a chosen path for the future [39].

The Other Side of F.A.I.R.: The FAIR principles allow researchers and scientists to do their jobs better, as well as being beneficial overall for the larger global society. However, when it comes to businesses, the likelihood of FAIR principles being incorporated is low, simply because data is vital to their services and products. There is little incentive to provide a framework to make data open and accessible to the wider population. But on the other side of the spectrum, a team at the MIT Media Lab has rethought current ideas around data management as well as storage. With the team presenting ideas that would transfer the personal data in the digital economy to be under the individual's control as they know how best to handle the risks that come with their data [15]). Researchers have found that if a dataset provides individuals' location hourly, and with spatial resolution that is equal to those provided by the mobile carrier's antenna, only 4 spatio-temporal points are needed to identify individuals 95% of the time. This reveals the gigantic dataset available and the relatively few number of points that are needed to track individuals' movements [14]. The researchers studied 1.5 million individuals' movements, but they argue that they can work for multiple population densities and geographical ranges [14]. Our movements are not as unique as we may believe them to be. And as a result, it is possibly even more important that individuals themselves hold on to their data [14].

OpenPDS/SafeAnswers: Researchers from MIT want users to have their data and decide who gets to use it. The idea would allow users to decide if the service gives them enough value for the data requested. This may create alternatives to the current business models of advertising and data-selling and can provide new business models focused on data collection or metadata storage [15]. As such, smaller companies may be able to enter the digital economy due to lower costs and thus can lead to a levelling of the playing field [15]. The approach, called "Safe Answers", would answer questions and would reveal data instead of returning metadata from users' phones [15]. The personal data store (PDS) that the user has would have the metadata. When asked, the researchers submitted a question in the form of code to the personal data store and the PDS would send back an answer also in the form of code [15]. So, the answer received by apps would lead to a reduction of traceable data. The only data transfer would be in code and questions, thus the problem of data anonymization would become a security problem. Users can also see what data is being shared with other apps on their phones [15].

More Arguments: Other ideas have also been presented, such as companies paying users for their data; consumers pay more to avoid data collection as well as targeted ads, and then consumers that allow these practices can receive a discount. [18]. However, some of these have been heavily criticized [51]. Another idea is that of personal data economy (PDE) that argues that companies should deal directly with the users, whose data is collected and advertised on their terms [18]. Many more ideas circulate in current discussions, nonetheless, what we do know is that, in the face of looming regulation, companies and users are facing a shift in the tides of the machination of the digital economy and of big data.

4.3   Use Case: Cookie Banners

In the last few decades, information consumption moved more and more online. Since many different people use the internet to consume information, the experience should be tailored to everyone. General news sites, for example, are meant to be used by a wide demographic [42]. Since 2018, websites have to get active consent from users for data collection [12]. This is in practice often done by using a cookie banner, to request consent from the user. However, as described in Sect. 3.3, sites often use nudging and dark patterns to guide users to certain decisions [27,32,42,52]. These so-called dark patterns [27] can lead to users consenting to things they do not fully understand. This would not be informed consent as is the goal of the GDPR [12], but is nevertheless still the de facto standard on many sites [52]. Several studies embraced this problem, e.g. according to Breen et al., the

> Current mechanisms for capturing consent urgently need redeveloping to be able to comply with the new regulation.

(in [7]). They pointed out, that there needs to be clear, simple and convenient mechanisms to request consent from data subjects. These mechanisms are currently often missing, as the user is often offered no choice regarding the degree of data collection [32]. While Sanchez-Rola et al. found the GDPR to have a big impact, they also noted that *"the spirit of the law is often still not applied"* (Sanchez-Rola et al. in [40]). This may be because of missing concrete rules on how to implement the GDPR. Specific guidelines, e.g. regarding cookies, are still missing [40].

Dark Patterns in Cookie Banners
Not only do many sites not offer any meaningful choice regarding cookie consent and data collection [16,52], but even if choices are offered, companies may use dark patterns to nudge users to their preferred decision [42]. This may be possible because of the GDPR's wording: *"the controller shall be able to demonstrate that the data subject has consented"* [12]. This does not prevent sites from using techniques to push users to a certain decision. For example, Utz et al. found 57.4% of consent notices use techniques like pre-selecting checkboxes and colour coding buttons [52]. These techniques lead the user to actions not necessarily in their own interest. Utz et al. furthermore note that if websites used banners

in accordance with the GDPR's principle of privacy by default, less than 0.1% of users would actively consent [52]. In practice, often more than 90% of users give their consent [32,52]. This discrepancy of theory and reality seems to indicate a structural problem with the GDPR, where its goals are not achieved by the rules it sets.

UI by Law

Based on the results found by Utz et al. [52], and in line with Nouwens et al. [32], Sanchez-Rola et al. [40], and Soe et al. [42], we propose an additional regulation to the GDPR, which specifies the user experience (UX) and the user interface (UI) design to be used by websites for compliance with the GDPR. This regulation could e.g. specify the cookie consent banner location (according to Utz et al., the bottom-left corner leads to the most attention of the user [52]). Furthermore, Nouwens et al. found that information below the first layer is effectively ignored by users [32]. Thus, the necessary information and UI would have to be placed on the first layer. Additionally, Utz et al. showed that pre-selection has a significant impact on users accepting or declining. Therefore, decisions regarding pre-selection have to be made. When taking the GDPRs principle of privacy first into account, only the absolutely necessary cookies should be pre-selected, nothing else. A very general UI example could look like Fig. 2. Regulations and specifications like this could help the GDPR to achieve in practice what was the goal in theory.

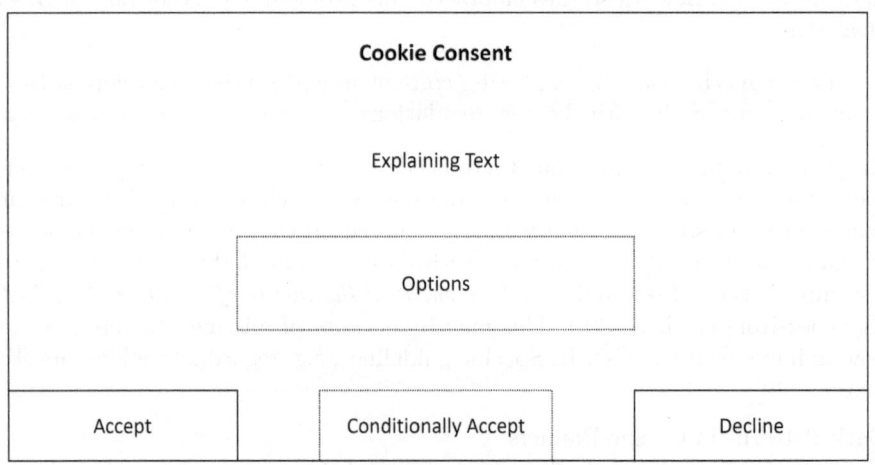

Fig. 2. Cookie Consent Example

Limitations in UI by Law

While the current situation is found problematic [32,40,52], and an UI regulation may address some of the problems, there are also other problems to be considered. For example, if the EU decides to further regulate the use of cookies, this may help with some issues regarding tracking via cookies. But that may also lead to companies using different techniques to identify users such as fingerprinting [31]. With fingerprinting, websites may use certain information about the user that they have, such as display size and resolution, to identify users independently of cookies. It has to be questioned, if regulation and legislation is able to keep up with new techniques developed by companies to track users.

5 Conclusion

This paper demonstrated some ethical, economic, and legal concerns concerning the current data economy. The answer to our research question, mentioned under the section "Introduction", is this: the novel pathways towards achieving an ethical data economy include an all-rounded approach that considers ethical issues from individuals', enterprises,' and legal perspectives. These novel pathways seek an immediate and collective strive by critical key stakeholders such as states, the EU, and (public/private) enterprises. For ethics assurance from an individual's perspective, immediate implementation of FACT (fair, accurate, confidential, and transparent) principles is required. Moreover, risk assessment of those measures, as per opacity, scale, and damage, is mandatory from the inception of (the technical and organizational) data handling measures. The economic impacts of data handling have a massive impact on companies' business models. As governments begin to step in with regulatory tactics, the data economy is rife for change. In the enterprise perspective, we analysed two possible data strategies that enterprises could adopt: FAIR (findable, accessible, interoperable, and reusable) principles or Open Personal Data Stores proposed by MIT Media Lab as per their respective business goals. The former approach asks for open platforms for data sharing across domains for the collective benefit of society and the latter demands that data subjects have an increase in control in sharing their personal information, in ensuring privacy preservation, and personal choice and control. The legal concerns regarding informed data collection include unclear cookie banners, which endanger the principle of privacy by default as set out in the GDPR. These problems are often rooted in dark patterns used by companies, which was also found by [16,42,52]. Thus, like Soe et al. [42] and Nouwens et al. [32], we propose further regulations regarding user interface design to address the goals of the GDPR. Furthermore, we give an example of how such regulation might look and show some possible limitations of the approach. Further research should be done to determine what such regulations could be exactly.

References

1. Un75 (2020). https://www.un.org/en/academic-impact/un75-social-contract-2020-toward-safety-security-sustainability-ai-world-0
2. Ec.ai (2022). @miscEC.AI.21, title=EC.AI21, https://digital-strategy.ec.europa.eu/en/library/proposal-regulation-laying-down-harmonised-rules-artificial-intelligence, year=2021

3. Rds (2022). @miscredasci.org, title=RDS, https://redasci.org/, year=2017
4. Aalst, W.M.P.: Responsible data science: using event data in a "People Friendly" manner. In: Hammoudi, S., Maciaszek, L.A., Missikoff, M.M., Camp, O., Cordeiro, J. (eds.) ICEIS 2016. LNBIP, vol. 291, pp. 3–28. Springer, Cham (2017). https://doi.org/10.1007/978-3-319-62386-3_1
5. Andrejevic, M.: Big data, big questions— the big data divide. Int. J. Commun. **8**, 17 (2014)
6. Bennett, S.C.: The right to be forgotten: reconciling EU and us perspectives. Berkeley J. Int'l L. **30**, 161 (2012)
7. Breen, S., Ouazzane, K., Patel, P.: GDPR: is your consent valid? Bus. Inf. Rev. **37**(1), 19–24 (2020)
8. Bulger, M., Taylor, G., Schroeder, R.: Data-driven business models: challenges and opportunities of big data. Oxford Internet Institute. Research Councils UK: NEMODE, New Economic Models in the Digital Economy (2014)
9. Burgess, M.: The tyranny of GDPR popups and the websites failing to adapt. Retrieved January 2022 **22** (2018)
10. Chan, K.: France fines google, facebook millions over tracking consent (2022). https://abcnews.go.com/International/wireStory/france-fines-google-facebook-millions-tracking-consent-82108667
11. Chiou, L., Tucker, C.: Search engines and data retention: implications for privacy and antitrust. Technical report, National Bureau of Economic Research (2017)
12. Council of European Union: Regulation (EU) 2016/679 (2016). https://eur-lex.europa.eu/legal-content/EN/TXT/PDF/?uri=CELEX:32016R0679&from=EN
13. Davis, K.: Ethics of Big Data: Balancing risk and innovation. "O'Reilly Media, Inc." (2012)
14. De Montjoye, Y.A., Hidalgo, C.A., Verleysen, M., Blondel, V.D.: Unique in the crowd: the privacy bounds of human mobility. Sci. Rep. **3**, 1–5 (2013). https://doi.org/10.1038/srep01376
15. De Montjoye, Y.A., Shmueli, E., Wang, S.S., Pentland, A.S.: OpenpDS: Protecting the privacy of metadata through SafeAnswers. PLoS ONE **9**(7), e98790 (2014)
16. Degeling, M., Utz, C., Lentzsch, C., Hosseini, H., Schaub, F., Holz, T.: We value your privacy... now take some cookies: measuring the GDPR's impact on web privacy. arXiv preprint arXiv:1808.05096 (2018)
17. Draper, N.A., Turow, J.: The corporate cultivation of digital resignation. New Media Soc. **21**(8), 1824–1839 (2019)
18. Elvy, S.A.: Paying for privacy and the personal data economy. Colum. L. Rev. **117**, 1369 (2017)
19. Fabiano, N.: Robotics, big data, ethics and data protection: a matter of approach. In: Aldinhas Ferreira, M.I., Silva Sequeira, J., Virk, G.S., Tokhi, M.O., Kadar, E.E. (eds.) Robotics and Well-Being. ISCASE, vol. 95, pp. 79–87. Springer, Cham (2019). https://doi.org/10.1007/978-3-030-12524-0_8
20. Goddard, M.: The EU general data protection regulation (GDPR): European regulation that has a global impact. Int. J. Mark. Res. **59**(6), 703–705 (2017)
21. Goldfarb, A., Tucker, C.E.: Privacy regulation and online advertising. Manage. Sci. **57**(1), 57–71 (2011)
22. Greze, B.: The extra-territorial enforcement of the GDPR: a genuine issue and the quest for alternatives. International Data Privacy Law (2019)
23. Hamilton, I.A.: (2021). https://www.businessinsider.com/apple-iphone-privacy-facebook-youtube-twitter-snap-lose-10-billion-2021-11?international=true&r=US&IR=T

24. Hartmann, P.M., Zaki, M., Feldmann, N., Neely, A.: Big data for big business? a taxonomy of data-driven business models used by start-up firms. Cambridge Service Alliance, pp. 1–29 (2014)

25. Herschel, R., Miori, V.M.: Ethics & big data. Technol. Soc. **49**, 31–36 (2017)

26. Janev, V., Graux, D., Jabeen, H., Sallinger, E. (eds.): Knowledge Graphs and Big Data Processing. LNCS, vol. 12072. Springer, Cham (2020). https://doi.org/10.1007/978-3-030-53199-7

27. Mathur, A., Kshirsagar, M., Mayer, J.: What makes a dark pattern... dark? design attributes, normative considerations, and measurement methods. In: Proceedings of the 2021 CHI Conference on Human Factors in Computing Systems, pp. 1–18 (2021)

28. McKendrick, J.: AI adoption skyrocketed over the last 18 months (2021). https://hbr.org/2021/09/ai-adoption-skyrocketed-over-the-last-18-months

29. Mehrnezhad, M.: A cross-platform evaluation of privacy notices and tracking practices. In: 2020 IEEE European Symposium on Security and Privacy Workshops (EuroS PW), pp. 97–106 (2020). https://doi.org/10.1109/EuroSPW51379.2020.00023

30. Morey, T., Forbath, T., Schoop, A.: Customer data: designing for transparency and trust. Harvard Bus. Rev. **93**(5), 96–105 (2015)

31. Nikiforakis, N., Kapravelos, A., Joosen, W., Kruegel, C., Piessens, F., Vigna, G.: Cookieless monster: Exploring the ecosystem of web-based device fingerprinting. In: 2013 IEEE Symposium on Security and Privacy, pp. 541–555. IEEE (2013)

32. Nouwens, M., Liccardi, I., Veale, M., Karger, D., Kagal, L.: Dark patterns after the gdpr: Scraping consent pop-ups and demonstrating their influence. In: Proceedings of the 2020 CHI Conference on Human Factors in Computing Systems, pp. 1–13 (2020)

33. O'neil, C.: Weapons of math destruction: How big data increases inequality and threatens democracy. Crown (2016)

34. Pery, A., Rafiei, M., Simon, M., van der Aalst, W.M.: Trustworthy artificial intelligence and process mining: challenges and opportunities. arXiv preprint arXiv:2110.02707 (2021)

35. Pirkova, E., Massé, E.: About Us - Access Now (2019). https://www.accessnow.org/eu-court-decides-on-two-major-right-to-be-forgotten-cases-there-are-no-winners-here/

36. Pollach, I.: A typology of communicative strategies in online privacy policies: ethics, power and informed consent. J. Bus. Ethics **62**(3), 221–235 (2005)

37. Popkova, E.G., Sergi, B.S.: Digital economy: Complexity and variety vs. rationality (2020)

38. Rantanen, M.M.: Towards ethical guidelines for fair data economy-thematic analysis of values of Europeans. In: Tethics, pp. 27–38 (2019)

39. Rios, R.S., Zheng, K.I., Zheng, M.H.: Data sharing during covid-19 pandemic: what to take away. Expert Rev. Gastroenterol. Hepatol. **14**(12), 1125–1130 (2020)

40. Sanchez-Rola, I., .: Can i opt out yet? GDPR and the global illusion of cookie control. In: Proceedings of the 2019 ACM Asia Conference on Computer and Communications Security, pp. 340–351 (2019)

41. Slobogin, C.: Principles of risk assessment: sentencing and policing. Ohio St. J. Crim. L. **15**, 583 (2017)

42. Soe, T.H., Nordberg, O.E., Guribye, F., Slavkovik, M.: Circumvention by design-dark patterns in cookie consent for online news outlets. In: Proceedings of the 11th Nordic Conference on Human-Computer Interaction: Shaping Experiences, Shaping Society. pp. 1–12 (2020)

43. Sohail, S.A., Bukhsh, F.A., van Keulen, M., Krabbe, J.G., Hruby, P.: Evaluating clinical-care metadata share and its fairification using the rea ontology. In: 16th International Workshop on Value Modelling and Business Ontologies, VMBO 2022. CEUR (2022)

44. Sohail, S.A., Allah, F., Krabbe, J.G.: Identifying materialized privacy claims of clinical-care metadata share using process-mining and rea ontology. In: 15th International Workshop on Value Modelling and Business Ontologies, VMBO 2021 (2021)

45. Sohail, S.A., Bukhsh, F.A., van Keulen, M.: Multilevel privacy assurance evaluation of healthcare metadata. Appl. Sci. **11**(22), 10686 (2021)

46. Sohail, S.A., Krabbe, J., de Alencar Silva, P., Bukhsh, F.A.: Privacy value modeling: A gateway to ethical big data handling. In: 14th International Workshop on Value Modelling and Business Ontologies, VMBO 2020 pp. 5–15. CEUR (2020)

47. Tavani, H.T.: Ethics, Computing, and Genomics: Edited by Herman T. Tavani. Jones & Bartlett Learning (2006)

48. Taylor, M.: How apple screwed facebook (2021). https://www.wired.co.uk/article/apple-ios14-facebook

49. Tene, O., Polonetsky, J.: Privacy in the age of big data: a time for big decisions. Stan. L. Rev. Online **64**, 63 (2011)

50. The European Commission: eprivacy regulation (2021). https://digital-strategy.ec.europa.eu/en/policies/eprivacy-regulation

51. Tsukayama, H.: Why getting paid for your data is a bad deal (2020). https://www.eff.org/deeplinks/2020/10/why-getting-paid-your-data-bad-deal

52. Utz, C., Degeling, M., Fahl, S., Schaub, F., Holz, T.: (un) informed consent: Studying gdpr consent notices in the field. In: Proceedings of the 2019 ACM SIGSAC Conference on Computer and Communications Security, pp. 973–990 (2019)

53. Vargo, S.L., Lusch, R.F.: The SAGE Handbook of Service-Dominant Logic. Sage (2018)

The Responsible Implementation of Artificial Intelligence in Childcare

R. N. Guérin[1], E. I. S. Hofmeijer[2] iD, L. M. Kester[3], and L. W. Sensmeier[4](✉)

[1] Leiden University, 2300, RA Leiden, The Netherlands
[2] Faculty of Electrical Engineering, Mathematics and Computer Science, University of Twente, 7500, AE Enschede, The Netherlands
e.i.s.hofmeijer@utwente.nl
[3] University of Twente, 7500, AE Enschede, The Netherlands
[4] University of Münster, Schlossplatz 2, 48149 Münster, Germany
sensmeier@uni-muenster.de

Abstract. Artificial Intelligence (AI) has become a bigger part of the daily lives of today's generation. However, it is still debated how AI applications affect children when they interact with them from an early age. Parents, children and society as a whole benefit when childcare practices lead to healthy outcomes later in life. However, for many (possible) AI applications in childcare, it is unclear if their effects promote positive outcomes for children or not. Therefore, the authors research the current degree of responsibility in the deployment of AI in childcare. Responsible AI applications should be safe, accepted, trusted, and closely aligned with human values, and in this case, ensure that the effects of AI on children's development are overwhelmingly positive. The authors give a broad overview of the current home, education, and healthcare AI applications while keeping the social and emotional effects in mind. To analyze if these AI applications are designed and implemented in a responsible way, the guidelines of Floridi et al. were used to investigate where improvements in current and future applications can be made. This assessment does not cover all human rights and ethical perspectives on AI. Nevertheless, it focuses on the responsibility of AI application implementation and in what way guardians, teachers, and medical practitioners should incorporate it into childcare.

1 Introduction

The question of how we should live with Artificial Intelligence (AI) systems has long been a part of our popular culture. Already in the 1950s, American writer Asimov wrote in his book 'I, Robot' how a human-like robot with artificial intelligence should be developed [8]. Back then, Asimov foresaw the potential problems that could arise between human-robot communication. He implemented 'Three

R. N. Guérin, E. I. S. Hofmeijer, L. M. Kester and L. W. Sensmeier—All authors contributed equally.

J. Bayer and C. Grimme (Eds.): AI, Human Rights and Ethics, LNAI 14400, pp. 113–133, 2025.
https://doi.org/10.1007/978-3-031-52082-2_8

laws'[1] that respect an AI system's moral and ethical implications and, therefore, their treatment of humans. To this day, Asimov's three laws are often still influential in books or films throughout science fiction and maybe even influence scientists who develop AI systems [8].

Since Asimov's time, the development of artificial intelligence has greatly advanced, and AI systems have become part of our daily lives. Especially for today's new generation, the role of AI systems is ever-increasing. This recent progress left the authors the question: in what ways are children affected by living with AI applications from an early age? AI has, for example, introduced new teaching systems and learning methods in the educational sector. In 2019, UNESCO published a working paper about the policy of AI education and expressed concerns about discrimination, limitations, and privacy [84]. Moreover, smart toys pose risks to children's security and privacy in a home and are, therefore, banned in Germany [50]. The rise of social robots used in the healthcare sector is also a cause for scrutiny [2]. The general use of AI systems raises considerable societal and ethical concerns. For example, most people have at least read something about an AI system that discriminated unfairly [21] or that AI made life-impacting decisions in a non-transparent way, as O'Neil wrote in her book 'Weapons of Math destruction' [58]. These impressions prompted the authors of this paper to investigate the state of AI applications in childcare and their (possible) effects on children. Therefore, we deduct the following research question for this paper: To which degree can AI systems in childcare be responsibly implemented, to ensure mainly positive effects from their interaction? To answer this question, a focus has been put on AI systems in the home setting, educational setting and healthcare. This scope was chosen to get a broad perspective of AI systems in childcare.

2   Background

In this part, we present necessary background information that we use to judge the responsibility of AI systems in childcare. This includes current looks on childcare, responsible AI, and current applications of AI technology in childcare.

2.1   Current Views on Childcare

One of the most important factors to judge the responsibility of an AI system is the expectations arising from the situation that they are used in. Therefore, in this section, we present the current perspectives on childcare as a basis for further analysis.

---

[1] First Law: A robot may not injure a human being, or, through inaction, allow a human being to come to harm. Second Law: A robot must obey orders given it by human beings, except where such orders would conflict with the First Law. Third Law: A robot must protect its own existence as long as such protection does not conflict with the First or Second Law.

What is seen as good childhood development, may be different per region and culture. While it is possible to generally define what kind of practices lead to healthy outcomes later in life, expectations vary; from aiding in providing for the family or with household chores, to playing and learning. This difference in expectations of children's behaviour causes a difference in the desired competencies children are expected to develop [14]. According to Parmar et al. the main factors determining what is seen as good childhood development by society are: "the physical and social settings of daily life, customs, and practices of care and parent psychology" [60].

However, at least three general goals in childhood development can be described; readiness to learn, success in school, and healthy outcomes later in life [7]. These goals are similar across different cultures, independent of what behaviour is expected from these children. They can be achieved by providing a stable environment that takes into account their health and nutritional needs, that provides safety, and promotes development and learning through stimulating and emotionally supportive interactions [13]. Parents or caregivers carry the responsibility to provide such a stable environment and its quality is determined within five areas; health, education, nutrition, child protection, and societal protection [13]. If a child's needs are met within each of these areas, the child will develop optimally with higher chances of success, independent of the kind of success [7].

There are many known ways to intervene in the factors that influence childhood development in order to improve the child's nurturing care. Literature suggests that successful and sustainable interventions should be built on existing healthcare and education platforms and should involve the children's different caregivers to provide the child with the level and type of care it needs throughout life [13]. Examples of these interventions are education about proper nutrition and providing high-quality pre-school education programs and materials. Some interventions in childhood development, including the use of AI systems, will be discussed in Sect. 2.3.

2.2 Responsible AI

Below, we attempt to define responsible AI, starting with defining AI as such. Following that, the understanding of responsible AI in literature is presented and questions to evaluate responsibility of AI systems are described. These questions will be used to assess responsible implementation of AI in Sect. 4.

The European Union describes artificial intelligence as the "ability of a machine to display human-like capabilities such as reasoning, learning, planning and creativity" [28]. Therefore, an AI system is a system that autonomously adapts its behaviour based on previous actions and their effects. AI systems can take the form of traditional software (e.g. virtual assistants or modern search engines) but also embodied forms like robots, autonomous cars, or drones [28]. As AI systems adapt autonomously and impact many areas of everyday life, the question arises if AI systems can act responsibly. Dignum describes Responsible AI as AI that is safe, accepted, and trusted. To this end, the goals, decisions, and actions of these systems have to be closely aligned with human values and expectations of the system [22]. A multitude of scholars and economic actors have created frameworks and decision aids for responsible AI in the past [3, 22, 41, 65, 92]. Floridi et al. developed the AI4People framework, which combines multiple commonly used frameworks for responsible AI [33, 34]. This framework consists of five core principles, which will each be addressed below.

The first principle, **beneficence**, is characterized as the promotion of well-being of humans and the planet. The well-being of humans does not only include physical and emotional integrity but also dignity. Beneficence is limited to the direct beneficiary and the environment affected by the system. From this principle, we derive the following question to evaluate if an AI system is beneficent:

Does the system promote Well-Being, preserve dignity, and sustain the planet?

The second principle is **non-maleficence**. AI systems should not only advance the common good, but the negative consequences of AI use also have to be considered. These negative consequences can arise accidentally or deliberately. Specifically, AI systems should promote privacy and security. Therefore, the following question has to be answered:

Does the system prevent a loss of privacy and security?

The third principle is **autonomy**, which is characterized by the power to decide. AI systems should be designed to not impair human freedom but promote it. Humans should therefore not be forced to delegate decisions to AI systems but rather be supported by it in their decision-making. From this, we deduct the following question:

Is the power to decide still in the hands of humans with this AI system?

The fourth principle is **justice**. Justice concerns the relationship of AI and citizens in society. Floridi et al. describe justice in three ways: AI should correct previous injustice, create shared benefits and prevent new injustices. An example of injustice could be unfair discrimination. Citizens should be able to thrive alongside AI, while AI systems should benefit society to build a better and more just future. From this we conclude the following question:

Does the system promote prosperity, preserve solidarity and avoid unfairness?

Floridi et al. name **explicability** as the fifth principle. This principle consists of two major parts. First, systems should not be opaque. A citizen that is affected by an AI system should be able to understand how this system works. The second part is accountability. Developers have to be accountable for the effects that their AI system has. From this principle, we conclude the following questions:

Does the system enable the previous principles through intelligibility and accountability?

2.3   Current Applications

To understand the responsibility of AI systems in childcare, we present current applications of AI systems in this section. Children can interact with AI systems on a daily basis. When they are at home, they can ask Alexa to put on their favourite song, can play AI-powered games, or their wearables can remind them that they have reached enough steps that day. They also might interact with platforms based on AI or even have a robot tutor at school. Additionally, children with special needs are sometimes diagnosed or are educated by AI-based software. A plethora of interaction possibilities between AI and children occur with both AI systems designed specifically for children, but also those that are not and children interact with anyhow. The following section will give a small overview of different interactions with AI in a child's life.

2.3.1   Home

At home, children come across most devices that are not specifically designed for them, even though versions that are adjusted for them might exist. An example of this is the Echo Dot smart home from Amazon. In 2018, they specifically released a version for children containing a rubber casing and having the ability for parental control [1,6]. The software is even being adapted to specifically pick up kids' voices, which are usually more high-pitched [40,73]. Other than that, it should also be able to deal with a child's developing language and therefore understand that "Alexa" could, for example, be pronounced as "Awexa". Apart from these non-child specific applications, there are also AI applications specifically designed for children, such as toys. Nowadays, children do not only interact with toys, but the toys also interact with them. These so-called smart toys can be placed in several categories, for instance, voice or image recognition toys [52], puzzles and building games [25], and health-tracking toys or wearables [9]. Some of these toys are smart devices that are not only used at home but also in an educational environment. The earlier mentioned Alexa, for instance, can also be equipped with educational games that aim to improve learning [30]. Additionally, toys, like My Keepon, are designed to be attentive and understand emotive actions [12]. Or Cozmo, where its AI has reactions eerily close to the way humans display emotion and represents itself as an adorable and trustworthy companion [68].

2.3.2   Education

The emergence of AI in education allows for multiple applications in both smart content and student engagement. Software platforms exist that focus on providing personalized education materials for areas like literacy and mathematics [47,48]. Such platforms may also use games or rewards to achieve better engagement with the subject [23,51]. For example, the Zenbo robot from the research by Weng et al. aims to promote self-regulated learning via robotic quiz games [90]. Surprisingly, AI systems are also developed to teach children about AI-related concepts and internet-of-things technology like the PopBots [91] and the

Any-Cubes [72]. More physical solutions can be found, for instance, in courses like Physical Education. In response to the COVID-19 pandemic, Shin et al. propose Jumple, a virtual physical education classroom. It is a remote learning environment that allows students to jump and play. Positions of body parts are detected by using AI-based pose estimation technology [76].

For some time, research has also focused on Intelligent Tutoring Systems (ITS). Such systems can provide customized instructions to its learners and aim at enhancing the learning experience of the individual [37]. This development was followed by Affective Tutoring Systems (ATS), which are supposed to take into account the affects or emotional states of the student [38]. An example of such a system was developed by Wang et al., who developed an emotional design tutoring system that aims to analyse their learning progress by collecting information on the user's emotions. When, for instance, negative emotions are recognized, this might indicate that the user is confused. To recognize the emotions, they use AI-driven facial emotion recognition and semantic text identification techniques [89].

2.3.3   Healthcare

In paediatrics, Socially Assistive Robots (SARs) are a common research topic. The preferred robot in human interaction research is the NAO robot, because of its open programming platform [2]. Applications of the robot are researched in various child healthcare areas, for instance emotional support [71], screening for autism in toddlers [70], or assessment, therapy and teaching of children with autism spectrum disorder [5,94], and possibly in paediatric asthma [29]. To optimally assist in such applications, AI-driven improvements are continuously being researched, like human emotion recognition from facial expressions [31,49], speech recognition [93], and automated planning for human-robot interactions [83]. Such applications are necessary to evaluate the child's emotional state and let the robot automatically respond accordingly. Additionally, Foster et al. aim to use the NAO robot to help children cope with painful procedures. In order to realize this, they want to incorporate social signal recognition to determine what state the children are in. Using a neural-network approach like Roffo et al. [69], they aim to automatically detect states, such as emotions, intention, and attitudes. Subsequently, they will use a similar AI-planning method as Petrick et al. [63], to decide on the appropriate response of the robot [35].

Besides SARs, AI is emerging in several different child healthcare areas. For young adults that suffer from anxiety or depression, a fully automated conversational agent has been designed to deliver cognitive behavioural therapy. The responses of the conversational agent were determined via an AI technique called decision trees [32]. Next to this, Balasuriya et al. aim to help visually impaired children to identify objects using AI and Computer Vision. Employing Artificial Neural Networks, objects in the scenery are recognized and, using a similar technique, a description of the identified objects is generated. Via a speech recognition module, the user is able to communicate with the system [10].

3   Effects of AI in Childcare

AI systems can have multiple effects on children, both positive and negative, that specifically stem from systems designed for children or more general systems. In this section, we present common impacts of the applications discussed before. These will act as a basis for the evaluation of responsible effects in Sect. 4.

Typical effects of AI systems in childcare might be specific for the field of childcare, or might also be more generally applicable. For example, effects following data protection and privacy affect every age group and are therefore more generally applicable. Legislation is already being set up and debated, with the aim to minimize the risks and allow for safe implementation [18, 24, 26, 27, 86]. A difference, however, might be that, in contrast to adults, children might be unaware of their data being collected. If children are aware that they are being monitored, it might be that they change their behavior [12].

Another effect that is important to highlight is the possible emergence of a digital divide. Disadvantaged communities might be similarly disadvantaged when the distribution or use of emerging AI technologies is concerned [85, 86]. If our world becomes increasingly more dependent on AI systems, some communities, and therefore their children, are disadvantaged in the digital world. Such consequences are profoundly important, but do not specifically result from the design of AI systems for childcare. Effects as a consequence of AI's specific design in childcare will be discussed below, by walking through the positive effects first, followed by contingent negative effects.

3.1   Positive Effects of AI in Childcare

Multiple positive effects of AI in childcare have been found that influence the emotional state, social skills, learning abilities and improve their well-being as a whole. Children might benefit from an emotionally stable companion. Parents could let their own desires influence those of their children. Or a teacher struggling with personal problems might overreact to children misbehaving or making a mistake. In the design of social robots, it might be possible to avoid such influences or reactions. The behaviour of robots is designed to be more consistent than that of humans and could therefore possibly be designed in such a way that it promotes healthy emotional development and better parenting techniques [20]. This could have a positive influence on children, since it can teach them how to behave in similar future situations. Borenstein et al. even argued that children could apply these techniques and behaviour in the future when they become parents themselves [12].

AI systems could also teach children to interact better with other children. Björling et al. found that the addition of a social robot in a group setting, promoted human-human interactions [11]. As an example, in a group setting, a (social) robot might call the name of a child who has been excluded from the group. This can result in the group making a place for the child to join and continue interactions together [44]. This can aid in both an ability to be more inclusive and a more inclusive environment as a whole. Furthermore, Rafique et al. hypothesized that it would be possible to teach children emotional intelligence while teaching them the fundamentals needed for leaning programming skills. At the end of their research, they found that both emotional intelligence and understanding of programming constructs were improved in children [67].

AI systems can also impact the well-being of children positively. Relationship-building behaviour of social robots can affect children's display of positive emotions. Interaction with robots as social agents could provide them with additional emotional support, similar to the feeling of belonging and acceptance a child might receive from a playmate [17]. For example, the 'Cozmo' robot uses AI to interact with children about their interests and is able to react to a child's emotions. Subsequently, children react to it in a similar fashion as they would react to a playmate [46]. Robots might also help children deal with stress originating from their environment or stress during hospitalization [11]. For example, if a child can engage with an AI powered virtual friend, they can forget about being in the hospital for a while. During hospitalization of children, several positive effects have been noted on children when six types of SARs were researched; less stress and pain, a distraction from pain or treatment, and in general more smiling, relaxation and openness [35,54,80]. Additionally, SARs are also used for children with autism spectrum disorder. Therapy for this is costly and time-consuming [66], and therefore it can benefit greatly from the application of SARs. Furthermore, the use of SARs might cause parents less stress, since it could relieve them of some duties if needed [78]. As for the children themselves, many studies have shown that SARs can increase the social engagement of these children with Autism Spectrum Disorder (ASD) and improve their social abilities [64,78,87].

AI systems can also have positive effects on formal education. Earlier studies had already shown that the use of Information and Communication Technology (ICT) in schools promotes collaboration and increases motivation through the availability of better information and shared work resources. This can result in a better understanding of the subject and increase creativity in communication [43]. Where traditionally, the whole class gets the same learning material, ICT can provide material adapted to the children's learning capabilities. Building on these developments, AI systems could raise the quality of education by making teaching and learning more personalized. This personalized teaching helps to increase children's engagement and through this, the learning experience [55]. The learning process can be made more interactive and make children more easily immersed in learning [4,39]. Additionally, AI systems in the classroom can also have an effect on teachers. These applications could grade work or could be active on discussion boards, which could allow the teacher to give more individual attention to the children [36].

3.2   Contingent Negative Effects of AI

Besides positive effects, we also found contingent negative effects. There exists one common effect most researchers and robotic developers are afraid of; children might anthropomorphize robots. This anthropomorphism might result in a different development of their empathy. Young children are more prone to such developments than adults because they are more eager for social connections and have less understanding of the workings of a robot. Anthropomorphism of robots may either be conscious or unconscious, of which the latter may also raise concerns about deception [74].

Children that spend a lot of time with AI systems could also develop impaired social and emotional skills. This in turn could lead to asocial behaviour or social isolation. Robots are encoded with a 'personality', and this may affect the emotional development of children. Robots do not have the ability to experience emotion in the present time genuinely. From this, the question arises if a robot can provide the level of care to children that is necessary. A child needs to see the other's emotional state when interacting, and a lack of showing empathy can therefore be criticized [79]. If children relate too consistently to robots, they might find that their love or friendship is not reciprocal. This might cause these children to lose the ability to have empathy [81].

On the other hand they might find robotic responses to be less judgmental than humans and therefore prefer their companionship [62]. Children might come to think that the robots are capable of being friends, and leaving children in total care of robots might cause an increase in the probability of psychological damage [74, 75, 82]. However, these kinds of negative effects are hard to prove since it is challenging to assess if empathetic development in children is really altered by social robots [61]. Some also argue that such effects will be prevented by developers of AI, since no company would want to be liable for damage to children. Therefore, the market would adapt to make responsible choices to reduce and prevent such effects [15]. Nonetheless, there exist real life examples that refute this kind of argument. For example, Facebook optimized their services for ads and screen time, even though this can cause psychological damage to children [16]. Although efforts are being made to hold them accountable for this, the process is slow and there is no guarantee for success. Even though this might be resolved in the end, the damage might already have been inflicted.

In addition to differences in emotional and social development, negative effects are also feared in the case of morality. Due to the difference between robot-human relationships and human-human relationships, children might not develop the same moral responsibilities as in human-human relationships [42]. Kubinyi et al. even make a comparison with cross-fostered animals. If such an animal is brought up by a surrogate species, it will behave differently. They fear this might happen when robots raise children as well [45]. Similarly, Melson argues that children brought up by robots will lack moral standing because they are prone to develop a different understanding of humans. This is due to the fact that they are not exposed to human morals in the same way children fully brought up by humans are [53]. One might make the case that AI is imitating humans and therefore will have a similar understanding. However, programming this might be too complex and therefore a much more simplified version is used.

On a different note, there is also a possibility of trauma by embodied robots. In a study by Coeckelbergh et al., it became evident that both parents and therapists prefer the robot to look like an animal or object, but not like a human. The authors discuss that it might be the case that they are afraid that a humanoid robot might replace human-human relationships. Additionally, they are anxious that the children might become too attached to a humanoid robot, and it may upset the children if they can no longer interact with it [19]. On the other hand, a child could also be traumatized by a human-like robot. A more human-like robot is appealing until a certain point; otherwise, a sensation of strangeness and eeriness could appear. This feeling is what is referred to as the 'uncanny valley sensation' [56]. This could lead to children becoming scared, similar to what happens when watching a movie that is not age appropriate.

Lastly, there could also be negative effects through goal conflicts between academic performance and overall development of children. AI systems in school, for instance, might focus on measurable academic skills instead of on established curricula. On the one hand, this performance-based focus could result in a decrease in motivation, contrary to the increase in motivation that can be gained from more personalized learning [57,59]. Furthermore, it could also lead to a neglect of skills that are less measurable but equally important.

4   Evaluation and Responsible Implementation

In this chapter, we are going to evaluate the responsibility of the AI applications described in Sect. 2.3 while looking at the home, educational and healthcare environment. As a basis for that, we are going to answer the questions about the responsibility of AI systems mentioned at the end of Sect. 2.2, relying on the positive and contingent negative effects described in Sect. 3.

Does the system promote Well-Being, preserve dignity, and sustain the planet?

When AI applications are used in the home environment, researchers worry that a lack of empathy shown by the AI system could lead to an underdeveloped emotional intelligence or even psychological damage. Many researchers warn that the use of AI systems in childcare in the home setting might affect children's social and moral development negatively [42]. However, it is also argued that the use of these AI systems could promote emotional development due to the fact that they can provide consistent companionship and act in the children's interests [20]. This is something that parents, teachers and other caregivers are not always able to provide, since they might have personal circumstances that limit them. So, if AI applications would be used to provide such stable companionship, in conjunction with parents and caregivers interacting with children to build their emotional intelligence, AI systems might be able to improve children's well-being while possible negative effects are minimized.

In the educational sector, the use of AI shows an increase in the quality of teaching, and thus can positively effect well-being. These applications show positive effects such as improved motivation and communication skills. Students are individuals and the educational sector could accommodate their specific needs and pre-existing strengths. AI technology can help to optimize the content and methods for each student through personalized learning. However, when AI systems fully replace teachers, the well-being of the student will suffer, since AI systems are not suited for teaching children about communicating knowledge and ideas. Furthermore, excessive reliance of the teacher on AI systems such as grade predictors might prompt them to give too little or too much attention to certain students, which can lead to a decrease in the quality of education [77]. This same effect can be seen in the students themselves; students could put in too little or too much effort in studying when they rely too much on AI systems like grade predictors. This all highlights the importance of the human factor, but also shows promise for improving education with the help of AI systems.

Lastly, AI systems in healthcare are specifically designed to improve well-being by helping children with disabilities or improving their comfort and happiness during hospitalization. Since these systems do not try to take over the roles of doctors and nurses, but focus on relieving stress or assistance in therapy, the well-being of users is improved. If AI systems continue to be designed with the goal of improving well-being by assisting caregivers, they could prove to be a beneficial addition to healthcare [66].

Does the system prevent a loss of privacy and security?

The protection of privacy and security is better preserved in the healthcare and education sector than in the home situation. In healthcare there are no specific indications that these applications would cause a loss of privacy, since the applications should be initiated by a therapist, doctor, parent or caregiver. In the education sector, privacy could even be promoted when using AI systems. Students need to communicate their thoughts with teachers, which causes an inherent loss of privacy. However, if these interactions take place with an AI system, their privacy might be preserved to a higher degree. A downside of this comes with the question if their data and usage of the system is being tracked. If this is the case, there is a huge amount of metadata that would otherwise not have existed that results in new privacy and security risks.

In the home situation such risks are larger, because AI systems often have little transparency in their data storage method and its accessibility. Enforcing these risks, is the fact that children, who will most likely be the main users, are probably less aware of the privacy and security concerns that are accompanied by the AI systems. In both the healthcare and educational setting, the main users are mostly their teachers or caregiver, who are more likely to be aware of the risks associated.

Is the power to decide still in the hands of humans with this AI system?

In both healthcare and education, it is clear that the power to decide is still in the hands of humans. In healthcare, AI systems are still only applied and researched in supportive roles. However, in both education and healthcare, the decision about using AI or not will impact the rights of patients and students as well. The possibility of opt-out for both patients and students or their caretakers is questionable. Further efforts are needed to elaborate on this before their deployment, to find a solution that respects the rights but also provides solutions to work with opt-outers.

Does the system promote prosperity, preserve solidarity and avoid unfairness?

The use of AI systems could promote prosperity and preserve solidarity in all three settings: healthcare, education and the home. The use of AI home systems can improve children's emotional intelligence, but can also teach them new skills like programming. Through this, AI systems can promote prosperity and solidarity indirectly, it might increase children's chances later in life, and make them more compassionate towards others. In education, AI systems could contribute to overcoming the inherent biases of teachers in assessing student performance, and possible not reproduce those biases, for example through grade prediction [77]. In addition, students may be less hesitant to interact with a software than with a teacher. These could result in improved education and therefore also an increase in prosperity. In the healthcare setting, a specific application like ASD can potentially promote solidarity, since it has been shown that it improves children's social skills.

To ensure fairness, specific and tailored measures must be taken in all three environments, which will further enhance the costs of the use of AI systems. Low-income households or lesser developed countries might not be able to afford such technology, increasing the risk of the possibility of a digital divide. It should be avoided that AI applications are made more affordable by reducing the ethical quality of the system in any phase of its lifecycle.

Does the system enable the previous principles through intelligibility and accountability?

Accountability and intelligibility are little discussed topics in all three settings. However, if AI systems are used in a supportive manner, the user could still be held accountable. This would therefore be feasible in the healthcare and educational setting, but could be questionable in the home setting. In all cases, however, it would nonetheless be imperative that the user has sufficient understanding and control over the used AI system, in order to be held accountable.

4.1 Responsible Implementation

Having evaluated the use of AI systems in the home, education, and healthcare setting using the question structure from Sect. 2.2, it can be clarified how these systems can responsibly be implemented. In all cases, but specifically in the home and educational setting, it is imperative that teachers and parents gain insight into the workings of the system. Since they would likely be held accountable for the AI system, they need to be able to make informed decisions on use and supervision. Furthermore, there are some slight differences in each setting, that could make implementation of AI systems responsible.

4.1.1   Home

AI systems could have serious negative effects on children's well-being as many researchers have discussed. Developers, of course, are interested in making their products safe, to avoid liability [15]. However, it has also been exemplified that such situations have occurred and are only being remedied after the damage has been done. To aid responsible implementation, parents should frequently be involved and be able to acquire sufficient knowledge about the applications. Providers of the AI systems should also be transparent about if and how data is acquired and stored. If this information is available, parents or guardians could decide to let their child use the AI systems or not. Accountability for the system would then be the responsibility of the parent or caregiver. Nonetheless, this highlights the need for understanding of the general working of the AI systems. This should therefore be realized in the application design.

4.1.2   Education

To responsibly implement educational AI systems, explanations and additional information about the system are required, specifically focused on teachers and students. It might even be convenient to educate teachers and students on AI technology. Education of AI technology in school for teachers and students is still in its developmental phase [88]. For this reason, it is unclear how experienced and educated teachers and students are on these applications. Teachers could misjudge their shortcomings and especially young students might also be unable to see the risks and benefits of AI systems in school. Next to education in AI technology, responsible implementation of AI systems should be in consultation with teachers and parents, as well as students. If every party is aware of the risks and benefits associated with the use of AI systems, one could prevent negative effects like bias of teachers and unmotivated students.

4.1.3   Healthcare

Outweighing the negative effects with the positive effects is largely in the hands of the users in the case of AI systems in healthcare. Most negative effects are centered around the fact that parents, but also therapists, do not want an AI system to fully replace a therapist or doctor. This can be prevented by designing the system in such a way that it will only support and not replace the user. This also immediately indicates that users are not limited by the system, since it has a supportive role, allowing the user to decide to use it or not. Ensuring this, child healthcare could greatly benefit from AI systems, since it can save parents and caretakers time and cause them less stress. Especially, children with disabilities or hospitalized children have shown improvements in their treatment and in their well-being, after interacting with an AI system.

5   Conclusion

In the introduction of this paper, the following research question was described:

To which degree can AI systems in childcare be responsibly implemented, to ensure mainly positive effects from their interaction?

To answer this research question, we divided the AI systems that a child interacts with into three groups. We see AI systems at home as not sufficiently responsible. Uses at home only offer debatable positive effects and potential downsides like diminished development of emotional intelligence or psychological damage. There also exist privacy and security risks. Moreover, the relations of control and accountability are unclear when more adults are using the system.

The second group of AI systems are systems in education. These systems offer largely positive effects such as leveraging pre-existing capabilities, and optimizing content for students. However, systems that predict students' performance or replace the teacher are negatively weighed. Furthermore, depending on the development, training, and use, AI systems can promote or demote fairness, so it depends on the application if there is a positive or negative effect on fairness. In addition, it is questionable if teachers and especially students are able to understand the consequences of using AI Systems.

Finally, in the field of healthcare, we found the least amount of negative effects. AI systems in healthcare can augment patient's care and therefore promote a patients' well-being. AI systems should however not replace practitioners so that the power to decide is still in a human hand. When using AI systems in childcare, it should be kept in mind that these systems are only tools. Therefore, the responsibility is dependent on how guardians, teachers, and medical practitioners incorporate it into childcare.

5.1   Limitations

This research aimed to provide a broad perspective of AI systems in childcare. This broad perspective comes at the cost of a less detailed analysis. The constraints on time and length did not allow us to take all available research into account. Furthermore, some of the discussed effects may be regarded as general consequences of AI systems, and therefore not necessarily specific to childcare, such as privacy risks. These general consequences are not discussed in depth, but do have an influence on the responsible implementation of AI systems in childcare.

The scope of this paper did not cover all ethical and/or human rights perspectives. Therefore, if an application is deemed responsible according to the framework used here, it could still be seen as unethical to implement, for other reasons. Lastly, this paper only considers the interpretation of responsible AI using the framework of Floridi et al. [34], which was selected to evaluate the AI application in this paper. Even though the chosen framework combines commonly used frameworks, is cited frequently, and is thus probably a well-accepted standard, the use of another framework might have led to slightly different results.

5.2   Outlook

Currently, many of AI's possible applications in childcare are not developed yet or are not applied on a broad scale. Suppose we, as a society, want to ensure responsible development of AI in childcare, frameworks, as the one used in this paper, should allow for the evolution of rules and technical standards. This would ensure continued responsible development of AI, as was also concluded by Floridi et al. If this is not done correctly, the development of AI for childcare is left to the market, which leaves the opportunity for companies to develop AI for their benefit instead of the benefit of society. However, if the development of AI in childcare is done following laws and standards that ensure responsible development, AI in childcare can assist parents, teachers and children in providing the best care and give children fairer chances to develop into their best selves. Therefore, it is important that we, the scientific community, continue the debate about responsible AI openly and inclusively [33].

References

1. Amazon Launching Echo Dot Kids Edition for Children — Time. https://time.com/5254163/amazon-echo-dot-kids-edition/
2. NAO the humanoid and programmable robot — SoftBank Robotics. https://www.softbankrobotics.com/emea/en/nao
3. ACT-IAC: ACT-IAC White Paper: Ethical Application of Artificial Intelligence Framework (2020). https://www.actiac.org/documents/act-iac-white-paper-ethical-application-ai-framework
4. Aktaruzzaman, M., Shamim, M.R., Clement, C.K.: Trends and issues to integrate ICT in teaching learning for the future world of education. Int. J. Eng. Technol. **11**(3), 114–119 (2011)
5. Alnajjar, F., Cappuccio, M., Renawi, A., Mubin, O., Loo, C.K.: Personalized robot interventions for autistic children: an automated methodology for attention assessment. Int. J. Soc. Robot. **13**(1), 67–82 (2021). https://doi.org/10.1007/s12369-020-00639-8
6. Amazon: Alexa for Kids - Learn how Alexa can help your family — Amazon.com https://www.amazon.com/alexa-for-kids/b?ie=UTF8&node=21474972011&ref=_alxhb_tpnv_kdfm
7. Anderson, L.M., et al.: The effectiveness of early childhood development programs. Am. J. Prevent. Med. **24**(3), 32–46 (2003). https://doi.org/10.1016/S0749-3797(02)00655-4
8. Asimov, I.: I, Robot. Gnome Press (1950)
9. Bagot, K., et al.: Current, future and potential use of mobile and wearable technologies and social media data in the ABCD study to increase understanding of contributors to child health. Dev. Cogn. Neurosci. **32**, 121–129 (2018). https://doi.org/10.1016/j.dcn.2018.03.008
10. Balasuriya, B.K., Lokuhettiarachchi, N.P., Ranasinghe, A.R.M.D.N., Shiwantha, K.D.C., Jayawardena, C.: Learning platform for visually impaired children through artificial intelligence and computer vision. In: 2017 11th International Conference on Software, Knowledge, Information Management and Applications (SKIMA), vol. 2017-Decem, pp. 1–7. IEEE (2017). https://doi.org/10.1109/SKIMA.2017.8294106

11. Björling, E.A., Thomas, K., Rose, E.J., Cakmak, M.: Exploring teens as robot operators, users and witnesses in the wild. Front. Robot. AI **7**, 1–15 (2020). https://doi.org/10.3389/frobt.2020.00005

12. Borenstein, J., Pearson, Y.: Companion robots and the emotional development of children. Law Innov. Technol. **5**(2), 172–189 (2013). https://doi.org/10.5235/17579961.5.2.172

13. Britto, P.R., et al.: Nurturing care: promoting early childhood development. Lancet **389**(10064), 91–102 (2017). https://doi.org/10.1016/S0140-6736(16)31390-3

14. Britto, P.R., Engle, P.L., Super, C.M.: Handbook of Early Childhood Development Research and Its Impact on Global Policy. Oxford University Press (2013). https://doi.org/10.1093/acprof:oso/9780199922994.001.0001

15. Bryson, J.J.: Why robot nannies probably won't do much psychological damage. Interact. Stud. Soc. Behav. Commun. Biol. Artif. Syst. **11**(2), 196–200 (2010). https://doi.org/10.1075/is.11.2.03bry

16. C-SPAN: Facebook Whistleblower Frances Haugen testifies before Senate Commerce Committee (2021). https://www.youtube.com/watch?v=GOnpVQnv5Cw

17. Cangelosi A, M, S.: Baby robots. In: Developmental Robotics. The MIT Press (2015). https://doi.org/10.7551/mitpress/9320.003.0005

18. CLAIRE: Response to the European Commission's Proposal for AI Regulation and 2021 Coordinated Plan on AI (August) (2021). https://claire-ai.org/wp-content/uploads/2021/08/CLAIRE-EC-AI-Regulation-Feedback.pdf

19. Coeckelbergh, M., et al.: A survey of expectations about the role of robots in robot-assisted therapy for children with ASD: ethical acceptability, trust, sociability, appearance, and attachment. Sci. Eng. Ethics **22**(1), 47–65 (2016). https://doi.org/10.1007/s11948-015-9649-x

20. Darwall, S.: Empathy, sympathy, care. Philos. Stud. **89**(2–3), 261–282 (1998). https://doi.org/10.1023/a:1004289113917

21. Dastin, J.: Amazon scraps secret AI recruiting tool that showed bias against women (2018). https://www.reuters.com/article/us-amazon-com-jobs-automation-insight-idUSKCN1MK08G

22. Dignum, V.: Responsible autonomy. In: Proceedings of the Twenty-Sixth International Joint Conference on Artificial Intelligence, vol. 0, pp. 4698–4704. International Joint Conferences on Artificial Intelligence Organization, California (2017). https://doi.org/10.24963/ijcai.2017/655

23. DreamBox: DreamBox Learning - Online Math Learning for Students, K-8. https://www.dreambox.com/

24. Ebers, M., Hoch, V.R.S., Rosenkranz, F., Ruschemeier, H., Steinrötter, B.: The European commission's proposal for an artificial intelligence act-a critical assessment by members of the robotics and AI law society (RAILS). J **4**(4), 589–603 (2021). https://doi.org/10.3390/j4040043

25. El Rhalibi, A., Wong, K.W., Price, M.: Artificial intelligence for computer games. Int. J. Comput. Games Technol. **2009**(1), 1–3 (2009). https://doi.org/10.1155/2009/251652

26. European Commision: White Paper On Artificial Intelligence - A European approach to excellence and trust. COM(2020), Brussels (2020). https://ec.europa.eu/info/sites/default/files/commission-white-paper-artificial-intelligence-feb2020_en.pdf

27. European Commission: Laying Down Harmonised Rules on Artificial Intelligence (AIA) and Amending Certain Union Legislative Acts. In: Regulation of the European Parliament and of the Council, vol. 0106 (COD). Brussels (2021)

28. European Parliament: What is artificial intelligence and how is it used? https:// www.europarl.europa.eu/news/en/headlines/society/20200827STO85804/what-is-artificial-intelligence-and-how-is-it-used
29. Ferrante, G., et al.: Social robots and therapeutic adherence: a new challenge in pediatric asthma? Paediatric Respiratory Rev. **40**, 46–51 (2021). https://doi.org/ 10.1016/j.prrv.2020.11.001
30. Filimon, M., Iftene, A., Trandabăţ, D.: Using games and smart devices to enhance learning geography and music history. In: Information Systems Development: Information Systems Beyond 2020 (ISD2019 Proceedings) (2019)
31. Filippini, C., Perpetuini, D., Cardone, D., Merla, A.: Improving human-robot interaction by enhancing NAO robot awareness of human facial expression. Sensors **21**(19), 6438 (2021). https://doi.org/10.3390/s21196438
32. Fitzpatrick, K.K., Darcy, A., Vierhile, M.: Delivering cognitive behavior therapy to young adults with symptoms of depression and anxiety using a fully automated conversational agent (Woebot): a randomized controlled trial. JMIR Mental Health **4**(2), e19 (2017). https://doi.org/10.2196/mental.7785
33. Floridi, L.: Ethics, Governance, and Policies in Artificial Intelligence. Springer, Cham (2021). http://www.springer.com/series/6459
34. Floridi, L., et al.: AI4People-an ethical framework for a good AI society: opportunities, risks, principles, and recommendations. Minds Mach. **28**(4), 689–707 (2018). https://doi.org/10.1007/s11023-018-9482-5
35. Foster, M.E., et al.: Using AI-enhanced social robots to improve children's healthcare experiences. In: Wagner, A.R., et al. (eds.) ICSR 2020. LNCS (LNAI), vol. 12483, pp. 542–553. Springer, Cham (2020). https://doi.org/10.1007/978-3-030-62056-1_45
36. Goel, A.K., Polepeddi, L.: Jill Watson: a virtual teaching assistant for online education. Georgia Inst. Technol. 1–21 (2016). https://fabricofdigitallife.com/Detail/ objects/3864
37. Graesser, A.C., Conley, M.W., Olney, A.: Intelligent tutoring systems. In: APA educational psychology handbook, Vol 3: application to learning and teaching, vol. 3, pp. 451–473. American Psychological Association, Washington (2012). https:// doi.org/10.1037/13275-018
38. Hasan, M.A., Noor, N.F.M., Rahman, S.S.B.A., Rahman, M.M.: The transition from intelligent to affective tutoring system: a review and open issues. IEEE Access **8**, 204612–204638 (2020). https://doi.org/10.1109/ACCESS.2020.3036990
39. Holstein, K., McLaren, B.M., Aleven, V.: Student learning benefits of a mixed-reality teacher awareness tool in AI-enhanced classrooms. In: Penstein Rosé, C., et al. (eds.) AIED 2018. LNCS (LNAI), vol. 10947, pp. 154–168. Springer, Cham (2018). https://doi.org/10.1007/978-3-319-93843-1_12
40. Huber, J.E., Stathopoulos, E.T., Curione, G.M., Ash, T.A., Johnson, K.: Formants of children, women, and men: the effects of vocal intensity variation. J. Acoust. Soc. Am. **106**(3), 1532–1542 (1999). https://doi.org/10.1121/1.427150
41. Jaume-Palasí, L., Spielkamp, M.: Ethics and algorithmic processes for decision making and decision support, AlgorithmWatch Working Paper No. 2, Berlin (2017)
42. Jones, R.A.: Representations of childcare robots as a controversial issue. Int. J. Mech. Aerospace Ind. Mechatron. Manuf. Eng. **11**(8), 1304–1308 (2017). https:// publications.waset.org/vol/128
43. Khan, M.S.H., Hasan, M., Clement, C.K.: Barriers to the introduction of ICT into education in developing countries: the example of Bangladesh. Int. J. Inst. **5**(2) (2012)

44. Komatsubara, T., Shiomi, M., Kaczmarek, T., Kanda, T., Ishiguro, H.: Estimating children's social status through their interaction activities in classrooms with a social robot. Int. J. Social Robotics **11**(1), 35–48 (2019). https://doi.org/10.1007/s12369-018-0474-7

45. Kubinyi, E., Pongrácz, P., Miklósi, A.: Can you kill a robot nanny? Interaction studies. Soc. Beh. Commun. Biol. Artif. Syst. **11**(2), 214–219 (2010). https://doi.org/10.1075/is.11.2.06kub

46. Kumar Singh, D., Sharma, S., Shukla, J., Eden, G.: Toy, tutor, peer, or pet? In: Companion of the 2020 ACM/IEEE International Conference on Human-Robot Interaction, New York, NY, USA, pp. 325–327. ACM (2020). https://doi.org/10.1145/3371382.3378315

47. Learning, C.: Literacy & ELA Solutions — Carnegie Learning. https://www.carnegielearning.com/solutions/literacy-ela/

48. Learning, C.: Math Solutions — Carnegie Learning. https://www.carnegielearning.com/solutions/math/

49. Lopez-Rincon, A.: Emotion recognition using facial expressions in children using the NAO robot. In: 2019 International Conference on Electronics, Communications and Computers (CONIELECOMP), pp. 146–153. IEEE (2019). https://doi.org/10.1109/CONIELECOMP.2019.8673111

50. Maras, M.: 4 Ways' Internet of Things' Toys Endanger Children (2018)

51. Math, T.: Online Math Tutoring and Coaching Programs — Thinkster Math. https://hellothinkster.com/online-math-tutoring-coaching-programs/

52. McReynolds, E., Hubbard, S., Lau, T., Saraf, A., Cakmak, M., Roesner, F.: Toys that listen. In: Proceedings of the 2017 CHI Conference on Human Factors in Computing Systems. vol. 2017-May, New York, NY, USA, pp. 5197–5207. ACM (2017). https://doi.org/10.1145/3025453.3025735

53. Melson, G.F.: Child development robots. Interact. Stud. Soc. Behav. Commun. Biol. Artif. Syst. **11**(2), 227–232 (2010). https://doi.org/10.1075/is.11.2.08mel

54. Moerman, C.J., van der Heide, L., Heerink, M.: Social robots to support children's well-being under medical treatment: a systematic state-of-the-art review. J. Child Health Care **23**(4), 596–612 (2019). https://doi.org/10.1177/1367493518803031

55. Mohd, C.K.N.C.K., Shahbodin, F.: Personalized learning environment (PLE) experience in the twenty-first century: review of the literature. In: Abraham, A., Muda, A.K., Choo, Y.-H. (eds.) Pattern Analysis, Intelligent Security and the Internet of Things. AISC, vol. 355, pp. 179–192. Springer, Cham (2015). https://doi.org/10.1007/978-3-319-17398-6_17

56. Mori, M., MacDorman, K., Kageki, N.: The Uncanny Valley [From the Field]. IEEE Robot. Autom. Magaz. **19**(2), 98–100 (2012). https://doi.org/10.1109/MRA.2012.2192811

57. Moseley, L.G., Mead, D.M.: Predicting who will drop out of nursing courses: a machine learning exercise. Nurse Educ. Today **28**(4), 469–475 (2008). https://doi.org/10.1016/j.nedt.2007.07.012

58. O'Neil, C.: Weapons of Math Destruction: How Big Data Increases Inequality and Threatens Democracy. Crown Publishers, USA (2016)

59. Owoc, M.L., Sawicka, A., Weichbroth, P.: Artificial intelligence technologies in education: benefits, challenges and strategies of implementation. In: Owoc, M.L., Pondel, M. (eds.) AI4KM 2019. IAICT, vol. 599, pp. 37–58. Springer, Cham (2021). https://doi.org/10.1007/978-3-030-85001-2_4

60. Parmar, P., Harkness, S., Super, C.M.: Asian and Euro-American parents' ethnotheories of play and learning: effects on preschool children's home routines and

school behaviour. Int. J. Behav. Dev. **28**(2), 97–104 (2004). https://doi.org/10.1080/01650250344000307

61. Pashevich, E.: Can communication with social robots influence how children develop empathy? Best-evidence synthesis. AI & SOCIETY (0123456789) (2021). https://doi.org/10.1007/s00146-021-01214-z

62. Pearson, Y., Borenstein, J.: The impact of robot companions on the moral development of children. In: Pirtle, Z., Tomblin, D., Madhavan, G. (eds.) Engineering and Philosophy. PET, vol. 37, pp. 237–248. Springer, Cham (2021). https://doi.org/10.1007/978-3-030-70099-7_12

63. Petrick, R.P.A., Foster, M.E.: Knowledge engineering and planning for social human–robot interaction: a case study. In: Vallati, M., Kitchin, D. (eds.) Knowledge Engineering Tools and Techniques for AI Planning, pp. 261–277. Springer, Cham (2020). https://doi.org/10.1007/978-3-030-38561-3_14

64. Pop, C., et al.: Can the social robot Probo help children with autism to identify situation-based emotions? A series of single case experiments. Int. J. Humanoid Robot. **10** (2013). https://doi.org/10.1142/S0219843613500254

65. PriceWaterhouseCoopers: PwC's Responsible AI: AI you can trust. PriceWaterhouseCoopers (2019). https://www.pwc.com/gx/en/issues/data-and-analytics/artificial-intelligence/what-is-responsible-ai/pwc-responsible-ai.pdf

66. Prior, M., Roberts, J., Rodger, S., Williams, K., Sutherland, R.: A Review of the Research to Identify the Most Effective Models of Practice in Early Intervention for Children with Autism Spectrum Disorders. Australian Government Department of Families, Community Services and Indegenous Affairs, Australia **01** (2011), https://www.dss.gov.au/sites/default/files/documents/10_2014/review_of_the_research_report_2011_0.pdf

67. Rafique, M., Hassan, M.A., Jaleel, A., Khalid, H., Bano, G.: A computation model for learning programming and emotional intelligence. IEEE Access **8**, 149616–149629 (2020). https://doi.org/10.1109/ACCESS.2020.3015533

68. Rahm-Skågeby, J.: "Well-behaved robots rarely make history": coactive technologies and partner relations. Des. Cult. **10** (2018). https://doi.org/10.1080/17547075.2018.1466567

69. Roffo, G., et al.: Automating the administration and analysis of psychiatric tests. In: Proceedings of the 2019 CHI Conference on Human Factors in Computing Systems, pp. 1–12, New York, NY, USA. ACM (2019). https://doi.org/10.1145/3290605.3300825

70. Romero-García, R., Martínez-Tomás, R., Pozo, P., de la Paz, F., Sarriá, E.: Q-CHAT-NAO: a robotic approach to autism screening in toddlers. J. Biomed. Inform. **118**, 103797 (2021). https://doi.org/10.1016/j.jbi.2021.103797

71. Rossi, S., et al.: Using the social robot NAO for emotional support to children at a pediatric emergency department: randomized clinical trial. J. Med. Internet Res. **24**(1), e29656 (2022). https://doi.org/10.2196/29656

72. Scheidt, A., Pulver, T.: Any-Cubes. In: Proceedings of Mensch und Computer 2019, New York, NY, USA, pp. 893–895. ACM (2019). https://doi.org/10.1145/3340764.3345375

73. Shahnawazuddin, S., Kumar, A., Kumar, V., Kumar, S., Ahmad, W.: Robust children's speech recognition in zero resource condition. Appl. Acoust. **185**, 108382 (2022). https://doi.org/10.1016/j.apacoust.2021.108382

74. Sharkey, A., Sharkey, N.: Children, the elderly, and interactive robots. IEEE Robot. Autom. Mag. **18**(1), 32–38 (2011). https://doi.org/10.1109/MRA.2010.940151

75. Sharkey, N., Sharkey, A.: The crying shame of robot nannies. Interact. Stud. Soc. Behav. Commun. Biol. Artif. Syst. **11**(2), 161–190 (2010). https://doi.org/10.1075/is.11.2.01sha

76. Shin, S., Cho, J., Kim, S.W.: Jumple: interactive contents for the virtual physical education classroom in the pandemic era. In: Augmented Humans Conference 2021, New York, NY, USA, pp. 268–270. ACM (2021). https://doi.org/10.1145/3458709.3458964

77. Starck, J.G., Riddle, T., Sinclair, S., Warikoo, N.: Teachers are people too: examining the racial bias of teachers compared to other American adults. Educ. Res. **49**(4), 273–284 (2020). https://doi.org/10.3102/0013189X20912758

78. Taheri, A., Meghdari, A., Alemi, M., Pouretemad, H.: Teaching music to children with autism: a social robotics challenge. Scientia Iranica **26**(1), 0–0 (2017). https://doi.org/10.24200/sci.2017.4608

79. Tronick, E., Adamson, L.B., Als, H., Brazelton, T.B.: Infant emotions in normal and pertubated interactions. In: Biennial Meeting of the Society for Research in Child Development, Denver, CO. vol. 28, pp. 66–104 (1975)

80. Trost, M.J., Ford, A.R., Kysh, L., Gold, J.I., Matarić, M.: Socially assistive robots for helping pediatric distress and pain. Clin. J. Pain **35**(5), 451–458 (2019). https://doi.org/10.1097/AJP.0000000000000688

81. Turkle, S.: There Will Never Be an Age of Artificial Intimacy (2018). https://www.nytimes.com/2018/08/11/opinion/there-will-never-be-an-age-of-artificial-intimacy.html?partner=rss&emc=rss

82. Turkle, S., Breazeal, C., Dasté, O., Scassellati, B.: First encounters with kismet and cog: children's relationship with humanoid robots (2006)

83. Umbrico, A., Cesta, A., Cortellessa, G., Orlandini, A.: A holistic approach to behavior adaptation for socially assistive robots. Int. J. Soc. Robot. **12**(3), 617–637 (2020). https://doi.org/10.1007/s12369-019-00617-9

84. Unesco: Artificial intelligence in education: challenges and opportunities for sustainable development. Working papers on education policy, 7 p. 46 (2019). https://en.unesco.org/themes/education-policy-

85. UNICEF: State of the Worlds Children 2017 - Children in a Digital World (2017). https://www.unicef.org/publications/index_101992.html

86. UNICEF: Policy guidance on AI for children (Draft 1.0) pp. 1–48 (2020). https://www.unicef.org/globalinsight/media/1171/file/UNICEF-Global-Insight-policy-guidance-AI-children-draft-1.0-2020.pdf

87. Vanderborght, B., et al.: Using the social robot Probo as a social story telling agent for children with ASD. Int. Stud. Soc. Beh. Commun. Biol. Artif. Syst. **13**(3), 348–372 (2012). https://doi.org/10.1075/is.13.3.02van

88. Vazhayil, A., Shetty, R., Bhavani, R.R., Akshay, N.: Focusing on teacher education to introduce AI in schools: perspectives and illustrative findings. In: 2019 IEEE Tenth International Conference on Technology for Education (T4E), pp. 71–77. IEEE (2019). https://doi.org/10.1109/T4E.2019.00021

89. Wang, C.H., Lin, H.C.K.: Emotional design tutoring system based on multimodal affective computing techniques. Int. J. Distance Educ. Technol. **16**(1), 103–117 (2018). https://doi.org/10.4018/IJDET.2018010106

90. Weng, T.S., Li, C.K., Hsu, M.H.: Development of Robotic Quiz Games for Self-Regulated Learning of Primary School Children. In: 2020 3rd Artificial Intelligence and Cloud Computing Conference, New York, NY, USA, pp. 58–62. No. 300, ACM (2020). https://doi.org/10.1145/3442536.3442546

91. Williams, R., Park, H.W., Breazeal, C.: A is for artificial intelligence. In: Proceedings of the 2019 CHI Conference on Human Factors in Computing Systems, New York, NY, USA, pp. 1–11. ACM (2019). https://doi.org/10.1145/3290605.3300677
92. Winfield, A.F., Michael, K., Pitt, J., Evers, V.: Machine ethics: the design and governance of ethical AI and autonomous systems [Scanning the Issue]. Proc. IEEE **107**(3), 509–517 (2019). https://doi.org/10.1109/JPROC.2019.2900622
93. Younis, H.A., Mohamed, A., Ab Wahab, M.N., Jamaludin, R., Salisu, S.: A new speech recognition model in a human-robot interaction scenario using NAO robot: proposal and preliminary model. In: 2021 International Conference on Communication & Information Technology (ICICT), pp. 215–220. IEEE (2021). https://doi.org/10.1109/ICICT52195.2021.9568457
94. Yousif, M.: Humanoid robot enhancing social and communication skills of autistic children: review. Artif. Intell. Robot. Dev. J. **1**(2), 80–92 (2021). https://doi.org/10.52098/airdj.202129

AI in Pre-employment Assessments; How Unbiased, Fair and Objective Is It?

Vasos Arnaoutis[1]([envelope]) [ID], Charlene Hinton[2] [ID], Lieke van Zijl[3],
and Leve Lorenzen[2]

[1] University of Twente, Enschede, The Netherlands
v.arnaoutis@utwente.nl
[2] University of Muenster, Münster, Germany
{chinton,l_lore03}@uni-muenster.de
[3] Leiden University, Leiden, The Netherlands

Abstract. Hiring processes have been in artificial intelligence and statistics. In the past years there has been an increasing number of digital solutions that can support the hiring processes. These new solutions claim objectivity and deduction of human biases from the recruiting process. This paper presents the current state of intelligent machine hiring systems by analysing two real use cases. From the use cases, the concerns and setbacks of the systems are highlighted showing that the technology need further growth before it can be considered impartial.

1 Introduction

Artificial intelligence (AI) has demonstrated the ability to vastly impact our society. With a diverse realm of applications changing industries, companies, governments, and the way of life, AI brings great benefits to humankind but also poses risks to human rights that can inflict negative outcomes. One of such risks concerns the right of humans to work [25].

More critically, the use of AI in HR has been shown to exacerbate existing biases in practice derived from historical data that culminate into discriminatory hiring practices. Such a case by e-commerce tech giant, Amazon.com Inc. has been the object of scrutiny by AI practitioners, policymakers, and researchers alike [10]. AI as applied in the field of human resources (HR) subjects job candidates to pre-employment screening, which inadvertently introduces certain biases in the selection process. Given that such a threat to human rights has already materialized in reality, it is, therefore, more crucial than ever to undertake the challenge of preventing future harms from occurring.

Thus, by employing a case study approach of companies specializing in the development of AI tools for HR, this paper seeks to answer the following research questions:

Are objectivity and fairness maintained in the assessment of potential job applicants even when using AI?

J. Bayer and C. Grimme (Eds.): AI, Human Rights and Ethics, LNAI 14400, pp. 134–145, 2025.
https://doi.org/10.1007/978-3-031-52082-2_9

How are objectivity and fairness demonstrated in pre-employment assessments?

In this way, we aim to contribute to research in the HR field by understanding how the existence of bias is mitigated in the use of AI in employee assessments. The remainder of this paper is structured as follows: Sect. 2 provides a brief background into the use of AI in HR as well as the challenges of its use throughout the hiring process. Section 3 explores two case studies selected for research involving video- and game-based AI pre-employment assessment tools. Section 4 discusses the objectivity and fairness of these tools critically placed in the debate surrounding bias. Lastly, Sect. 5 concludes the paper and suggests further avenues of research.

2   Background

2.1   AI in HR

The OECD defines an artificial intelligence (AI) system as a "machine-based system that can, for a given set of human-defined objectives, make predictions, recommendations, or decisions influencing real or virtual environments" [20]. The benefits that artificial intelligence (AI) promises have spurred development and application in various areas of society. One of these benefits is the increase in operational efficiency as more and more organizations leverage their data to power algorithms that help run business operations and improve performance. In the human resources (HR) department where workforce planning and learning and development are core activities, the use of AI can be observed to analyse employee performance, determine efficient work schedules, and allocate organizational resources [9,11]. Furthermore, AI in HR has shown to widen the diversity of the pool of job candidates as well as reduce recruitment costs [7]. They are also perceived to achieve fairness by removing the implicit, unconscious bias that clouds human judgment in hiring decisions [25]. A recent report indicated that about 17% of organizations use AI in HR, and this number is expected to climb to 30% by 2022 [9].

2.2   Challenges of the Use of AI in HR

While the benefits of using AI are significant, HR departments face a host of challenges as they embark on the journey of applying AI in their organizations. The first challenge relates to the complexity of HR outcomes, particularly what is defined as a "good employee" and how to measure the factors influencing this [26]. These researchers assert that distinguishing between individual, team, and organizational performance that contribute to employee outcomes is difficult. On top of this, identifying the scope of performance indicators and then measuring them precisely is challenging. Secondly, HR outcomes such as firing an employee do not happen frequently in smaller organizations to provide sufficient training data for algorithms. And lastly, there remains ambiguity in the legal frameworks regarding who is accountable for HR decisions made by AI [26]. Drawn from the United Nation's Declaration on Human Rights (UDHR), these challenges affect job applicants' "Five Human Rights" such as the human right to work, quality and non-discrimination, privacy, free expression, and free association, which highlight the importance of addressing these questions [25].

2.3   Bias and Fairness

Researchers discuss the negative facet of AI application in HR concerns the effects of implicit, negative bias and discrimination that AI introduces when it comes to hiring practices. Because these AI systems make decisions that influence large groups of people, real-life examples of its negative effects have surfaced in the form of discrimination, or the unfair or unequal treatment of people or groups based on characteristics such as race, gender, or socio-economic factors [8]. For example, Amazon's hiring algorithm was discovered to favour men over women due to the training dataset used, which represented mostly men in the past decade [10].

In her book, Caroline Criado Perez mentions how even job vacancies that are not explicitly mentioning sex, can unconsciously use terms that are more tailored to men. Similarly, such word selection can favour different categorizations and clusters, as an example; extroverts. It has been shown that job vacancies, in some countries, can be more tailored towards extroverts or introverts, causing imbalance in the hiring percentage between the categories [8].

Ferrer et al. suggest that while bias is associated with discrimination in AI, bias does not lead to discrimination (2021). They define bias as "a deviation from the standard," the presence of which is sometimes needed to identify statistical patterns in data. Interestingly, the American Psychological Association (APA) argues that differences in outcomes do not indicate bias but that the significant presence of group differences should indicate sources of potential bias [2]. Researchers have identified that the main, problematic causes for bias in AI are bias in modeling, training, and usage. However, they caution that eliminating bias in AI systems does not guarantee a non-discriminatory system, but that it depends on the context in which the system is being used [8].

While no universal definition exists for fairness, fairness is "the absence of any prejudice or favouritism towards an individual or group based on their intrinsic or acquired traits in the context of decision-making" [16]. In ML, the concept of fairness generally translates to equal odds, opportunities, consistent similar outcomes regardless of groups, and so forth but fulfilling these parameters of fairness turns out to be challenging [16].

Explainability, or how well criteria used in AI-decision making are understood, is viewed as one of the ways of establishing fairness in the successful use of AI [26]. Indeed, algorithms perform better than human judgment when used to predict repetitive outcomes. Yet when tasked with explaining these outcomes, particularly in life-changing matters such as diagnosing cancer or hiring employees, experts reject the actionable recommendations of the AI because understanding how the AI came to its decisions became difficult [26]. Previous research has shown that humans trust algorithmic decisions less than human decisions [15,19]. Inadvertently, the final decision on important matters is left with the biased human [7].

2.4 AI-Facilitated Hiring Process

The hiring process is composed of four stages: sourcing, screening, interviewing, and selection [24]. In each stage, organizations have used AI tools to facilitate the process of finding talent and assessing fit through pre-employment screening, while also expanding diversity and inclusion. These tools can be in the form of chatbots, social media behaviour pattern recognition, speech recognition, facial analysis, serious game simulations, background checks, and offer negotiations. However, recent research indicates that these tools tip the power dynamics in favour of the employer and not the job applicant. For example, using a robot as a proxy in employment interviews, researchers Norskov et al. found that applicants perceived robot-mediated interviews to be less fair than face-to-face human interviews [19]. They suggest that this intervention may be seen as a way to control the process rather than to inject fairness in the hiring decision (2020, p. 14).

Similarly, video interviews provide flexibility and cost reduction in the hiring process but are also associated with scepticism and negative perception by applicants [4]. This power imbalance pits the job applicant against the AI tools that they may not fully understand or even know about the employer, while the employer knows significantly more than them. To this extent, AI-facilitated hiring processes introduce harm to the applicant due to lack of transparency and disadvantaged power dynamics. Despite the benefits that AI presents in providing fairness in the hiring process by reducing human biases, the very same tools may run the risk of worsening the very conditions it sought to alleviate under the guise of objectivity [7].

3 Case Study: Video-Based Assessment

In the screening stage of the hiring process (previously mentioned in Chap. 2), game- and video-based assessments in particular are increasingly common [24]. A form of video-based assessment is the video interview analysis. This is a form of pre-employment assessment that algorithmically evaluates candidates on the basis of video interviews [24]. There are several companies that provide this kind of assessment as a service. Relatively well-known examples are HireVue, Modern Hire and Knockri, which are treated in more detail in this chapter.

In these video interviews, depending on the company's application in question, candidates are either asked to record answers to certain questions or open-ended video resumes where they talk about their strengths (and pitfalls). Afterwards, the videos are then analysed by the respective AI models of the company, which differ for each of the vendors of the video analysis software [24]. The candidates' response can be analysed in terms of content, word choice, intonation, and micro facial movements, which are facets that can be associated with a company's successful employee [6]. The reason video interview analysis is becoming more popular is that pre-screenings are often costly, time-consuming, and biased. So, it saves time for the recruiter, which can be put into other essential tasks, and the major advantage is that it reduces bias and thus discrimination. Another benefit is that the software is developed in such a way to assess the candidate that fits best with the job and the organization [1].

Nevertheless, AI-based models are not perfect, and thus video screening software is no exception to that. The AI is as good as the data that you feed it. There are several flaws that should be kept in mind when using AI in the hiring process, namely inherent bias and other ethical concerns such as what data and how data is actually being analysed. In this chapter, three vendors of interview screening software will be researched (via their website) in terms of how their software (models) work, and how they treat bias and fairness in their systems. First, the most popular company for screening software, HireVue, is being discussed. Then Modern Hire and lastly Knockri are treated.

3.1   HireVue

As previously mentioned, HireVue is one of the most well-known companies that offer video interview screening software. They create job-specific algorithms or models to analyse the data from the video interviews. Big companies such as Unilever and Goldman Sachs are known to use HireVue's software.

HireVue's video interview application focuses on a structured and consistent interview; asking the same questions, in the same manner, and evaluating the response against the same job-related criteria [12]. The video interview software notices everything and analyses what is said and how someone said it. The AI evaluates the "hard" and "soft" competencies. The first being the skills that have something to do with the contents or calibre of the job, while the latter is more about communication skills and problem-solving abilities. These competencies are based on research by occupational psychologists who figured out what characteristics and skills are needed or important for a specific job role [13]. So, the only data points that are said to be used in the model are the ones that help predict success in the job [14]. The previously described structure provides ground for the two primary agenda points of HireVue: 'increasing diversity' and 'mitigating bias'. They increase diversity (by 16%) by being able to consider more applicants and to evaluate a wider pool of candidates because the process is mostly automated, which reduces the time to hire (by 90%) [12]. The other major aim, 'mitigating bias', is being pursued by actively and methodically auditing the algorithms to filter out the possibility that it is not adversely impacting[1] a certain group of candidates. If the results appear to be biased, HireVue investigates ways to remove the problematic features, and then retrains the model to address the adverse impact [14].

---

[1] Adverse impact is the negative effect of unfair and biased selection on a protected group, which occurs when this group is discriminated against during, for example, a hiring process [18].

On their website, HireVue also states to follow the Equal Employment Opportunity Commission (EEOC) Uniform Guidelines when developing, testing, and monitoring their AI-driven assessments. The EEOC Guidelines are in place to forbid employment practices that discriminate against race, colour, sex, religion and national origin [5]. In Sect. 4, Subsection D, adverse impact is being countered by the $\frac{4}{5}$ rule, also known as the Red-Flag Rule. This rule states that "if the selection rate for a certain group is less than 80% of that of the group with the highest selection rate, there is an adverse impact on that group" [18]. Furthermore, HireVue says to, next to following the EEOC Guidelines and the $\frac{4}{5}$ rule, use "other statistical tests for group differences", which they use to monitor and mitigate their assessments [18].

To summarize, HireVue reduces bias by using only the data points that are valid and necessary to predict someone's success in a potential job. The other thing they do is controlling and checking their algorithms for bias against certain candidates, and then removing the features that cause bias. Besides, HireVue also follows the guidelines of the EEOC and uses the $\frac{4}{5}$ rule to counter adverse impact.

3.2 Modern Hire

Another platform for video interview software is Modern Hire (previously Montage). Modern Hire has different interview technologies, such as live video and phone interviews, automated interview scheduling and AI-enabled Automated Interview Scoring [17]. The latter is being discussed in this section. The Automated Interview Scoring software is used in video interviews. It is said to fasten the hiring process, mitigate bias, and increase efficiency and fairness. The AI technology analyses the candidate's interview responses and suggests scores based on job-relevant data and key competencies associated with job success. The candidates are ranked according to their overall score, after which the best candidates can be selected to move forward in the hiring process. The AI analysis is said to only focus on the content of the answers, and not on the image, facial expressions, or audio qualities of the candidate in question. Modern Hire also states that their Automated Interview Scoring ensures an increase in objectivity and accuracy, since the scoring system is a consistent and standardized process. It is said to have more than three times less interview bias. Also, their AI models analyze 100.000 candidate responses to verify prediction, audit adverse impact, and reduce interview bias. Lastly, the model is trained on 1.5 million interview responses to understand language in the context of an interview [17]. The scoring feature described above, is part of CognitIOn (by Modern Hire). CognitIOn is Modern Hire's industry-leading science, where they combine the principles of industrial-organizational (I-O) psychology with the practical application of advanced artificial intelligence (AI) techniques, and billions of candidate interactions. They specifically mention that CognitIOn uses deep learning (a form of machine learning) and other statistical techniques to create their software. The website also mentions that Modern Hire offers an ethical 'Glass-Box Science' approach, which provides transparency into the data; how it is collected, evaluated and used. In addition, Modern Hire does not analyse facial features or social media profiles [17].

Furthermore, Modern Hire considers three standards for ethical AI, which are transparency, verifiability, and reproducibility. The first, transparency, refers to the data behind the employment decisions of the model. This means, it should be clear, how the factors that determine these decisions are evaluated and used. The second, verifiability, means that claims by an AI-based product should be backed up by explanations of the data; how it is collected, analysed and what it predicts. The last, reproducibility, is about the fact that AI researchers and developers should describe their methods openly and share their findings with others [17].

To summarize, Modern Hire works with an Automated Scoring System, which is part of CognitIOn. This system is said to mitigate bias and increase efficiency and fairness. The AI only analyses the contents of the candidate's responses, and not facial features and other sensitive data. More importantly, Modern Hire has three standards for ethical AI: transparency, verifiability and reproducibility. These standards are very important for addressing bias and fairness in Modern Hire's process.

4   Case Study: Game-Based Assessment

As previously mentioned, another assessment method that is used in the screening stage and implements AI is game-based. In short, candidates are asked by companies to do "behavioural-based exercises" [22] to find, among other things, candidates who come from a non-traditional background or who have not had the usual academic trajectory. The AI is used in the evaluation procedure to prevent a possible bias. There are several providers of game-based assessments, such as Pymetrics or perspect ai [22].

4.1   Pymetrics

Pymetrics game-based assessment approach is illustrated by several case studies, that include Accenture, McDonald's or Mastercard, among others. After submitting an online application, candidates are then invited to participate in Pymetrics exercises. In only 25 min, the participants are asked to complete 12 engaging games to measure emotional and cognitive attributes accurately and fairly, whereby data is permanently collected and evaluated. Furthermore, logical, and numerical reasoning can be tested by adding four further tests. After the games have been completed by the candidates, the accrued data is processed by the AI and a personal report is created for each candidate. In this personal report, the emotional, social, and cognitive skills are assessed, thus predicting the match with the company, which has previously created a success profile. According to a case study done with an innovative law firm, 24% of the final offers were given to non-traditional candidates [22], indicating the results of Pymetrics' implementation. Pymetrics claims to be able to measure attention, effort, fairness, decision making, emotion, focus, generosity, learning, risk tolerance, quantitative reasoning, and numerical agility, while also being able to reduce bias and increase fairness in the assessment process. The latter is ensured by several factors. First, according to Wilson 2021, no demographic characteristics, or "overt proxies for demographic information" (p. 7), such as ZIP codes, were used during model training. Furthermore, the seven Equal Employment Opportunity Commission (EEOC) groups were used (male, female, white, black, Hispanic, Asian and people who identify with more than two ethnic or racial groups [28]). Furthermore, Pymetrics claim in their compliance whitepaper, that they extract behavioural traits from the games and examine them "for their potential to cause gender and racial/ethnic bias using statistical methods" [23] in order to create an unbiased AI that goes beyond the requirements of Uniform Guidelines on Employee Selection Procedures (UGESP).

4.2   Perspect AI

Just like Pymetrics, Perspect AI also uses cognitive tests to gather important information about candidates, such as, their ability to solve problems, think and reason, while also providing accurate and fair measures of emotional, cognitive, and behavioural attributes [21]. In detail, when a job role is created, the design assessment takes place where the desired traits and skills are defined. The subsequent application process is similar to Pymetrics and mainly involves games. At the end, an insightful report is created on each candidate, where the company can base its decision on. In this report, you can find information on problem solving abilities, attention, information processing, behaviours, and also personality traits and emotional intelligence. Overall, Perspect AI claims to take the test takers' attention away from the fact that they are under evaluation, analysing their true behaviour, while also reducing social desirability bias. However, there is not much information about how exactly Perspect AI works on reducing its bias.

4.3   Arctic Shores

Another game-based assessment provider is Arctic Shores. Just like the previously mentioned providers, Arctic Shores uses game-like elements and interactive tasks to get information on a candidate's personality and ability, while also providing a fair and unbiased opportunity for applicants. According to Arctic Shores, the classical application process that includes a CV makes it impossible to hire applicants fairly. Reasons for this are that a lack of work experience of graduates means that recruiters have to focus on subjective information due to a lack of useful data. Furthermore, they say that already the name of the applicant is crucial for the success of the application [3]. As a solution, Arctic Shores developed a behaviour-based assessment to guarantee a fast, effective, and fair hiring process. Subjective data is to be replaced by objective data and each candidate is to be assessed individually and his or her suitability for the role in the company is to be evaluated. For this assessment, age, gender, background, neurotype and ethnicity are not considered.

5   Discussion

This section will critically discuss the findings from the study cases presented in the paper. The goal is to subjectively judge the level of objectivity and fairness of the current AI hiring practices. The structure of the analysis is broken down into two components; transition of bias from human to machine and back, and poor use of AI tools for the sake of technology advancement.

5.1   Shifting the Bias

Multiple papers, organizations, and software companies will refer to their solutions as AI-based systems. In many cases, the term is used to describe statistics, machine learning or language processing solutions. The presence of AI is still questioned by experts in the field, who feel that the term is misused. The reality for the current AI solutions, is that either statistical or learning based, they all learn from data. There are two inputs that can influence the performance of these algorithms; the dataset and the programmer. On the basis of this, claims of unbiased algorithms cannot be made, as in its core, an AI solution is built by two biased sources [27].

Nevertheless, biases are not necessarily negative and that not all biases lead to discriminatory practices [8] In the domain of AI, well-thought and desirable biases can be considered as "features", as long as these features fit the purpose of the solution. How can we then separate and define what are good biases? Within our presented literature, authors will describe an established AI solution as "efficient and effective" when it fulfills their hypothesis and definition of objectiveness [6]. This, for example, can be through measuring diversity within the hiring group; an increase in women hired is a representation of objectivity in the selection of candidates within an AI-hiring agent. As mentioned by Perez, identifying fault-lines, such as minimizing the use of terms tailored to a broad set of attributes, can introduce desired biases and evaluation metrics towards the AI algorithms that can improve the fairness of the hiring process.

Are these objective criteria stable? Or are they dependent on the societal acceptable and current trends? Pymetrics has shown that the evaluation of their AI solution has shown an increase in diversity between seven ethnic groups as defined by the EEOC. This clustering, however, can be subject to change or can be considered a discriminating measure as the society grows. The state of society has matured in the past years, by acknowledging and accepting people under different definitions. This suggests that a hypothesis for evaluating an AI tool in the present can be deemed incomplete in the future, which would mean that biases in the current solutions can knowingly or unknowingly exist.

Considering the core of building current AI solutions, their evaluation criteria and hypothesis for objectiveness, it is fair to say that a claim of an unbiased

AI hiring tool cannot be made without defining these limitations. However, considering the previous practices of human-centric hiring processes, a comparison between that and the AI-hiring tools can show a relative improvement and balance between recognition of various groups which can favour the decision making process of the AI.

5.2   Use of AI for the Sake of AI

On the other hand, there are some claims opposing the involvement of AI in the hiring process. As shown from the use cases, the establishment of an AI component as part of the hiring process may not yield positive results either by the reaction of the candidates or due to the uncertainty of the application. Such example is the implementation of a facial analysis as part of a hiring process from HireVue, which was later deemed unnecessary and was retracted. These attempts of AI integration have been characterized by literature as means to prove technological advancement, innovation and novelty, which in some cases has backfired. While for most practices the integration of AI shows systematic improvements, it is important to keep in mind that not everything should or is required to be substituted by automated machines and doing this prematurely can have bad influence on the specific organization and also in the technology's reputation itself.

Similarly, current AI-hiring systems have an end to end communication with an HR expert, meaning that the representative of the company will set up the guidelines for the vacancy and the requirements, and once the AI has shortlisted candidates, it will be up to the expert to make the final decision. While this in itself can be considered necessary, for the close monitoring of the AI solution itself, it also suggests that the human bias remains part of the system. While the role of the human is reduced, the critical and final decision can still be influenced by the embedded biases of the decider. Given the trade-off between keeping the algorithm or the human accountable for the final recruitment decision, it could be argued that the use of AI is doomed to be limited within the employee hiring process, thus limiting its capabilities towards real objectivity and fairness.

In conclusion, the benefits of selective use of AI in the hiring process can be apparent and rewarding, both for the company but also for the applying candidates. Collectively, the current rewards from the implementation of AI, both in reducing the hiring process lead time but also towards building a more manageable and controllable set of biases, is an indication that AI could benefit the HR domain, under close supervision and control. At its current state, both the AI but also the understanding and expression of the fairness towards hiring new recruits needs continues iterations and maturity growth.

References

1. Albert, E.T.: AI in talent acquisition: a review of AI-applications used in recruitment and selection. Strateg. HR Rev. **18**(5), 215–221 (2019). https://doi.org/10.1108/SHR-04-2019-0024

2. American Psychological Association: Principles for the validation and use of personnel selection procedures — fifth edition. https://www.apa.org/ed/accreditation/about/policies/personnel-selection-procedures.pdf

3. Arctic Shores: How the cv stops you hiring your grads fairly — arctic shores, 22 December 2021. https://www.arcticshores.com/insights/cvs-fair-hiring-diversity/

4. Basch, J., Melchers, K.: Fair and flexible?! explanations can improve applicant reactions toward asynchronous video interviews. Pers. Assess. Decis. **5**(3) (2019). https://doi.org/10.25035/pad.2019.03.002

5. Biddle Consulting Group: EEOC uniform employee selection guidelines questions and answers (2015). https://www.uniformguidelines.com/uniformguidelines.html#18

6. Black, J.S., van Esch, P.: AI-enabled recruiting: what is it and how should a manager use it? Bus. Horiz. **63**(2), 215–226 (2020). https://doi.org/10.1016/j.bushor.2019.12.001

7. Chamorro-Premuzic, T., Akhtar, R.: Should companies use AI to assess job candidates? (2019). https://hbr.org/2019/05/should-companies-use-ai-to-assess-job-candidates

8. Ferrer, X., van Nuenen, T., Such, J.M., Cote, M., Criado, N.: Bias and discrimination in AI: a cross-disciplinary perspective. IEEE Technol. Soc. Mag. **40**(2), 72–80 (2021). https://doi.org/10.1109/MTS.2021.3056293

9. Gartner: AI shows value and gains traction in HR, 19 January 2022. https://www.gartner.com/smarterwithgartner/ai-shows-value-and-gains-traction-in-hr

10. Guardian: Amazon ditched AI recruiting tool that favored men for technical jobs. The Guardian, 11 October 2018. https://www.theguardian.com/technology/2018/oct/10/amazon-hiring-ai-gender-bias-recruiting-engine

11. Harvard Business Review: What does building a fair AI really entail? (2020). https://hbr.org/2020/09/what-does-building-a-fair-ai-really-entail

12. HireVue: Structured interview software — hireVue hiring platform (2022). https://www.hirevue.com/platform/online-video-interviewing-software/structured-interview-builder

13. HireVue Team: How to prepare for your HireVue assessment, 16 April 2019. https://www.hirevue.com/blog/candidates/how-to-prepare-for-your-hirevue-assessment

14. Larsen, L.: Reducing bias and widening the candidate pool: why we built HireVue assessments, 24 May 2018. https://www.hirevue.com/blog/hiring/reducing-bias-and-widening-the-candidate-pool-why-we-built-hirevue-assessments

15. Lee, M.K., Rich, K.: Who is included in human perceptions of AI?: Trust and perceived fairness around healthcare AI and cultural mistrust. In: Kitamura, Y. (ed.) Proceedings of the 2021 CHI Conference on Human Factors in Computing Systems, pp. 1–14. ACM Digital Library, Association for Computing Machinery, New York (2021). https://doi.org/10.1145/3411764.3445570

16. Mehrabi, N., Morstatter, F., Saxena, N., Lerman, K., Galstyan, A.: A survey on bias and fairness in machine learning. ACM Comput. Surv. **54**(6), 1–35 (2021). https://doi.org/10.1145/3457607

17. Modern Hire: Video interviewing — virtual interview — modern hire, 01 December 2021. https://modernhire.com/platform/interview/
18. Mondragon, N.: Creating AI-driven pre-hire assessments **2021**, 07 June 2021. https://www.hirevue.com/blog/hiring/creating-ai-driven-pre-employment-assessments
19. Nørskov, S., Damholdt, M.F., Ulhøi, J.P., Jensen, M.B., Ess, C., Seibt, J.: Applicant fairness perceptions of a robot-mediated job interview: a video vignette-based experimental survey. Front. Robot. AI **7**, 586263 (2020). https://doi.org/10.3389/frobt.2020.586263
20. OECD: RBC-and-artificial-intelligence. https://mneguidelines.oecd.org/RBC-and-artificial-intelligence.pdf
21. Perspect AI: Cognitive tests, 20 October 2021. https://perspect.ai/solutions/talent-acquisition/cognitive-tests-and-simulations.html
22. Pymetrics (2022). https://www.pymetrics.ai/
23. Pymetrics Whitepaper: Compliance with EEOC guidelines
24. Raghavan, M., Barocas, S., Kleinberg, J., Levy, K.: Mitigating bias in algorithmic hiring. In: Hildebrandt, M. (ed.) Proceedings of the 2020 Conference on Fairness, Accountability, and Transparency, pp. 469–481. ACM Digital Library, Association for Computing Machinery, New York (2020). https://doi.org/10.1145/3351095.3372828
25. Skorburg, J.A., Yam, J.: Is there an app for that?: Ethical issues in the digital mental health response to COVID-19. AJOB Neurosci., 1–14 (2021). https://doi.org/10.1080/21507740.2021.1918284
26. Tambe, P., Cappelli, P., Yakubovich, V.: Artificial intelligence in human resources management: challenges and a path forward. Calif. Manage. Rev. **61**(4), 15–42 (2019). https://doi.org/10.1177/0008125619867910
27. West, S.M., Whittaker, M., Crawford, K.: Discriminating systems: gender, race and power in AI (2019). https://ainowinstitute.org/discriminatingsystems.html
28. Wilson, C., et al.: Building and auditing fair algorithms. In: Proceedings of the 2021 ACM Conference on Fairness, Accountability, and Transparency, pp. 666–677. ACM Digital Library, Association for Computing Machinery, New York (2021). https://doi.org/10.1145/3442188.3445928. https://evijit.github.io/docs/pymetrics_audit_FAccT.pdf

AI-Controlled Autonomous Tractors in German Agriculture

Kian Deutz and Dominik Neumann[(✉)]

Department of Information Systems, Westfälische Wilhelms-Universität Münster,
Münster, Germany
{kian.deutz,d_neum11}@uni-muenster.de

Abstract. This paper examines the challenges surrounding the implementation of autonomous tractors in German agriculture. Germany, a pioneer in precision agriculture, has yet to fully utilize autonomous driving technology for tractors due to legal, technical, and ethical obstacles. While advanced autonomous tractors are being developed, legal barriers and concerns about job losses and social inequalities hinder their adoption. Future research should explore these issues through a structured literature review. To unlock the potential of autonomous tractors in improving productivity and efficiency, German institutions must address ethical concerns and adapt policies to maintain their leadership in digital agriculture.

1 Introduction

To a great extent, the agricultural industry is responsible for ensuring food security, an ever-increasing challenge, with a growing world population expected to reach ten billion by 2050 [4,13]. This challenge combines a reduced available labour workforce and a more 'corporate' style of farming [3, p. 499]. Therefore, new technologies are being developed and deployed to overcome these challenges and increase efficiency and productivity. In this sense, German agriculture is portrayed as a driver of digitalization and has already taken a pioneering role in the adaptation of digital technologies years ago and has maintained this until today, so that already around 82% of German farms use smart farming technologies [13,15].

Artificial Intelligence (AI), as part of these emerging technologies, is seen as having great potential in agriculture, as it is particularly well suited to problems that cannot be solved well by humans or traditional computer structures. In a dynamic structure where situations cannot be generalized to propose a general solution, AI can grasp each situation and offer solutions that are best suited to the problem at hand [1]. A large application area for AI is autonomous driving, which in terms of agriculture means the autonomous driving and working of tractors and machines. These can then work the fields with little or no human intervention, which brings several advantages in terms of efficiency and is therefore seen as the future and desired state of agriculture. Furthermore, there is a

J. Bayer and C. Grimme (Eds.): AI, Human Rights and Ethics, LNAI 14400, pp. 146–155, 2025.
https://doi.org/10.1007/978-3-031-52082-2_10

labour shortage in the field of agriculture, which can be filled with autonomous driving agricultural machinery. Since agricultural work is highly seasonal and time-sensitive, autonomous systems fit perfectly for the case. During harvest season, every minute of harvesting the field counts before the rain comes and destroys the harvest. Autonomous systems can work day and night since they do not need certain working hours, which leads to less idle time. Despite these advantages and the fact that Germany is considered a role model in terms of novel technologies in agriculture, autonomous tractors are not yet used in Germany, which brings the objective of this paper to life. It is mainly the legal hurdles that have prevented their application so far and thus slowed down technological progress, although today's technology is quite ready for such an implementation. In addition, there are technical requirements and ethical hurdles to the introduction of autonomous tractors. This leads to the following research question of this paper: What are the technical requirements as well as the legal and ethical obstacles for autonomous driving of tractors in German agriculture? In order to answer this question, the following chapter first provides basic knowledge on the topic of AI and autonomous driving, after which details on the technical realization are presented. This is followed by a discussion of ethical and legal hurdles, of which the discussion in the last chapter focuses primarily on the question of whether autonomous tractors could be used in Germany in the future.

2   Fundamentals

According to Farhud and Zokaei, "Artificial intelligence (AI) is a term applied to a machine or software and refers to its capability of simulating intelligent human behaviour, instantaneous calculations, problem-solving, and evaluation of new data based on previously assessed data" [5, p. 1]. AI is being used extensively, particularly in the context of autonomous driving, as its main role is to improve driving safety while minimizing the effort required of the human driver [14]. Autonomous driving is divided into five different levels in the literature [6], based on the classification system of the Society of Autonomous Engineers (SAE) International. According to Goldfain et al. [6] and Jiru [8], level 0 vehicles are those that are under the full control of the driver. Level 1 includes driver assistance services such as adaptive cruise control, while level 2 is called partial automation, where the vehicle can perform some manufacturer-determined safety measures on its own, including emergency braking, parking assistance, and autopilot. At level 3, conditional automation, the highly automated vehicles can drive automatically under certain conditions by monitoring the environment, but the human driver must still be able to take back control of the vehicle if the autonomous system fails. At level 4, the high automation, the vehicle safely takes control and can proceed accordingly if there is no response to the request for human intervention. At this level, the driver can therefore sleep, although driving in unmapped areas or in unsafe weather conditions is not recommended. At level 5, full automation, the vehicles are fully automated under all conditions and in all operating modes and can handle extremely complex driving situations. At this level, vehicles can even operate without passengers [6,8]. In the following, only autonomous tractors are considered that correspond to level 5, i.e., they drive and operate fully autonomously and do not require human assistance, except for the monitoring of the vehicle by workers. AI is used in various application areas of agriculture. Today, three main artificial intelligence techniques are used to tackle agricultural problems: Expert systems, artificial neural networks, and fuzzy systems [1]. Application areas include general crop management, disease management, pest management, agricultural product monitoring and storage control, weed management, yield prediction, and soil and irrigation management.

In connection with the increasing automation of agricultural processes, the term "precision agriculture" is also frequently used. In precision agriculture, also known as smart farming [13], new technologies are used to increase crop yields and profitability while reducing the number of traditional inputs needed to grow the crops [17]. Precision agriculture leads to an overall increase in the investments and resources needed to grow the crops, using technologies such as online weather forecasts, sensor systems, networks, satellite systems, but also artificial intelligence [2]. In this sense, the application area that is the focus of this work is that AI is used to enable the autonomous driving of tractors or other agricultural vehicles and thus contribute to increasing efficiency in the sense of precision agriculture. According to Wessling [21], the market for autonomous tractors in the global agricultural robotics market, which is expected to grow from USD 4.9 billion in 2020 to USD 11.9 billion in 2026, is expected to surpass drones and milking robots by 2024. Despite the expected growing market share and the fact that Germany is considered a pioneer in terms of new technologies in agriculture, autonomous tractors are not yet to be found in the field today, which is why the technical requirements, as well as ethical and legal hurdles, are presented below.

3   Technical Realization

The Technical Realization of autonomous tractors can be split into two different fields. On the one hand, a tractor needs certain hardware to be able to perform autonomous tasks, and on the other hand, it needs software that handles the decision process instead of a driver.

On the hardware side, there are different sensors that enable autonomous driving, pathfinding, and hazard detection. These include but are not limited to optical sensors, electro-optical sensors, ultrasonic sensors, and LIDAR for image tasks, as well as GPS and cellular connectivity for positional tracking as well as internet communication. In general, sensors are split into pose and object sensors. Pose sensors locate where the tractor is, whereas object sensors detect hazards or other relevant objects surrounding the tractor. Sensors can be active or passive: active sensors emit energy, while passive sensors just detect information. Since all sensor types have their unique advantages and disadvantages, a combination of sensors is required [19]. In addition to the different sensors of an agricultural machine, the information received by them needs to be processed. A performant computing unit needs to be part of the tractor if the calculations are done on-site and not cloud-based.

Current tractors, like the current version of the TRION 700 of Claas already have most of the hardware necessary for performing autonomous agricultural tasks[1]. However, there is no fully autonomous tractor on the market yet in Germany (July 2022). The tractors already have the hardware since it is used for solving an agricultural routing problem (ARP). Creating the optimal path is crucial for optimizing field efficiency and utilizing high-end technology to reduce operating costs based on a potential 16% reduction in travel distance compared to a conventional broadcast treatment [7].

One promising way for achieving autonomous tractor control is through the utilization of deep learning technology. Deep learning is a form of artificial intelligence that involves creating a model with a specific input matrix, such as an image, and a neural network that processes the information and produces an output matrix. An example of this is object detection, where the goal is to locate a specific object within an image. In this scenario, the image serves as the input, and the output consists of four coordinates defining the bounding box of the target object. Deep learning models are trained using labeled training data. During training, the image is processed by the neural network, and if the network does not produce the desired output specified by the label for the given input image, the adjustable internal parameters of the network are modified through backpropagation to minimize the error. The network gets better the more training data it receives [11]. The advantages of deep learning in autonomous navigation of agricultural machinery are that it can drive the vehicle in real-time without prior knowledge of the terrain. Furthermore, it offers the crucial advantage of GPS-only navigation systems which detect hazardous objects in the path of the vehicle. It can detect the crop edge in the path and adjust the vehicle in order to minimize the path it needs to harvest the whole field [9]. With the current status of technology, it is possible to have an AI-controlled autonomous tractor or combine for agricultural tasks. The only hurdle to the realization of these systems is the amount of data needed in order to build a neural network capable of controlling the machines, tractors, or combines.

4   Barriers

This chapter looks at the ethical and legal aspects of the use of AI in general, each with reference to the focus of this article: autonomous tractors.

4.1   Ethical

AI is a hotly debated topic in science and practice for a reason, and not only because of the ethical hurdles that arise from its application. AI was therefore already very controversially advertised in various films decades ago, almost

---

[1] https://www.claas.de/produkte/maehdrescher/trion/models.

underpinned with a dystopian vision. This goes hand in hand, above all, with people's fear that robotic artificial intelligence will maliciously take on a life of its own, turning from a supportive role into a threatening one for humans. Nowadays, the actual negative aspects and risks are minimal compared to this dystopian idea to reassure humanity but should not be underestimated and should therefore be strictly pursued. Farhud and Zokaei [5] examine the medical and healthcare industry in terms of ethical issues, grouping them into issues of privacy and protection, social gaps and justice, informed consent and autonomy, medical consultation, empathy, and compassion. Although the focus is on a different industry sector, some of these issues are also considered relevant to agriculture.

The social gap created by AI is particularly challenging. Although AI has the potential to improve accessibility, it exacerbates social inequality by widening the gap between developing and advanced countries [5]. This challenge also arises with autonomous tractors, as the initial costs are incredibly high due to the highly developed technology and can only be financed by large companies, which tend to be located in industrialized countries. Moreover, this aspect also reinforces the problem that smaller farms do not stand a chance against large companies, as they are unlikely to be able to bear the price of these tractors on their own. On the contrast it could also be an opportunity for third world countries if they have access to autonomous agricultural machines. The problems of poverty and food insecurity can be reduced by raising the agricultural productivity of a country [20]. Mark [12] also explores the problem of inequality between farmers in his study. In his study, discovering journals of ethical issues in agriculture related to the application of AI and Big Data, he notes that there is a digital divide between farmers who are able to adopt such new technologies and those who are not. In general, small farms outnumber large farms worldwide, with the majority of agricultural production taking place on small farms with little technology. As technologies are mainly used in large industrial monoculture farms, as the use of such technologies is expensive and poorer farmers are discouraged from adopting them, this could lead to a disproportionate growth of larger farms and the dissolution of smaller farms [12].

Another much-discussed argument against the use of AI and the associated increase in automation is the replacement of human labor with machines [5]. Many farmers who use a certain number of workers on their fields could reduce this number enormously by using an autonomous tractor, as only a smaller number of workers are needed to control, maintain and repair these vehicles. This leads to many workers losing their jobs, thus increasing the unemployment rate. From the employer's point of view, however, the total amount of wages paid may remain the same, even though the number of workers is reduced since highly qualified personnel are now needed, who must obviously be better paid.

Another important issue is the attitude of farmers towards the use of AI. Although AI has great research potential, the attitude and acceptance of farmers must also be present in order to use AI and thus realize its full potential. Mohr and Kühl [13] have explored this topic in greater depth and searched for factors that promote and increase the acceptance of the use of AI in agriculture. Above all, perceived ease of use plays a major role, which is why the system must be user-friendly, for example. In addition, the farmer's personal attitude must be present, which primarily reflects his own interest in improving progress through AI. Finally, perceived benefits are also an important pillar for the acceptance of AI in agriculture, which is why the benefits its use promises should always be advertised. Rübcke von Veltheim and Heise [18] also investigated farmers' attitudes towards the introduction of autonomous field robots. They interviewed 490 German farmers and found that they are generally open to new technologies but want to understand the concrete benefits before adoption, which again underlines the importance of properly promoting the benefits.

4.2   Legal

The legal barriers to autonomously working farming machinery can be split up into two regards. First, the machinery needs to be able to go to the field and back from the field autonomously. Second, the machinery needs to perform a certain task on the field autonomously.

For driving to the field with public infrastructure, the same rules apply to any autonomous vehicle. In Germany, a fully automated driving function is allowed as long as this function can be deactivated or overtaken at any time (compare §1 (a) para. 2 Nr. 3 StVG). However, this law presupposes that there is a person sitting in a driver's seat, which would not be the case for fully autonomous agricultural machinery. The German law considers only level 4 automation [10], whereas, level 5 automation is needed. In this regard, it is legally not possible to perform the full task autonomously.

On the field itself, other regulations apply. The ISO standard 18497 gives principles for automated agrarian machines in the field. It has six core principles that need to be fulfilled. These 6 principles are (according to the "Gesetzliche Vorgaben für den Feldeinsatz von hochautomatisierten Landmaschinen 2018"):

1. The machine needs a perception system that can scan and process the environment and can localize persons and other obstacles.
2. A localization system needs to localize the machine and restrict it from leaving a work area.
3. Before starting, it needs to be made sure that there is no person or obstacle in the dangerous area of the machinery.
4. If there is an obstacle in the dangerous area of the machinery during the automated process, the machine must play an alarm, and it needs to be switched to a safe state, defined by the manufacturer.
5. The person using the machine needs to be able to start and stop the machine on site and remotely.
6. The machine needs to be able to be surveiled remotely.

Principles 1, 2, 3, 5, and 6 are feasible for the current state of autonomous machinery. Principle 4 could stop the machine from doing farm work. However, since it can be switched on and off remotely based on principle five, it can resume working if the danger is gone. For doing the actual farm work autonomously, there are no legal boundaries that restrict a way that it is not feasible to farm autonomously.

5   Introducing John Deere 8R Tractor

The US American company John Deere is considered a forerunner for technologically innovative products in the agricultural sector. In addition to innovations such as autonomous field sprayers and electric tractors, they have also developed the John Deere 8R autonomous tractor, depicted in Fig. 1.

Fig. 1. John Deere 8R Tractor, Source Image: Auto Motor Sport (2022)

Figure 1 John Deere 8R Tractor, Source Image: Auto Motor Sport (2022)
   This tractor is regarded as one of the most promising autonomous tractors and, according to the manufacturer's own information, is not only a prototype but is ready for serial production and thus available on the market in 2022. This tractor is equipped with six stereo cameras and monitors the autonomous tractor's surroundings 360°, using AI to support decision-making by analyzing the images, detecting obstacles, and calculating distance [16]. According to Wessling [21], the John Deere 8R tractor drives completely autonomously so that the workers can occupy themselves with other tasks. An app simply allows the tractor to be monitored on the smartphone. This is met with enthusiasm by some farmers. First, this helps to fill in the significant labour shortage in agriculture, a mostly seasonal work, and enables workers to do two tasks at the same

time. Second, rather than taking away jobs, it even creates new areas of responsibility and thus employment, by raising the needs for monitoring, control, and maintenance of the tractors. The attitude of workers towards the new technology represented by autonomous tractors is regarded as positive, due to the simplification of work processes. Nevertheless, it is important to keep in mind that the usage of the systems should be as simple as possible. However, John Deere announced that the market launch would be only for a limited amount of North American customers. Furthermore, autonomous tractors will not be launched in Europe since safety regulations do not permit the use of autonomous vehicles[2]. For Germany, particularly, this is the case because of §1 (a) para. 2 No. 3 StVG, which requires a person sitting in the driver's seat. This can only allow level 4 automation.

6 Conclusion

The starting point for this work was the increasing importance of new technologies in the agricultural sector, with Germany as a pioneer and role model for precision agriculture. Artificial intelligence that imitates human behaviour is used to enable autonomous driving of vehicles, which can be divided into five different stages. Autonomous tractors of the fifth stage are already far advanced in their development but are not yet in use, at least in Germany. This is mainly due to legal obstacles raised by German legislation. In addition, ethical concerns regarding labour losses and social inequalities are increasingly being discussed.

Due to its limited scope, this paper could not include a detailed methodological approach, such as a structured literature review, and is therefore not representative of all research and practice. Future research could therefore shed more light on academic literature with the help of a structured approach, and present the current state of autonomous tractors and their design. In recent years, more and more companies have made it their business strategy to bring autonomous tractors for farms onto the market, but in the European market, ethical considerations and national legislation hinder the introduction of such autonomous tractors. To realise the full potential that such tractors promise in terms of improved productivity and efficiency, and thus potentially meet rising food demand, it would be justified for national governments, specifically the German institutions to find their way to address the ethical concerns and alter their policies to remain a pioneer in digital agriculture in the future.

---

[2] https://www.deere.de/de/unser-unternehmen/news-und-medien/
pressemeldungen/2022/january/vollig-autonomer-tractor-8r410.html.

References

1. Bannerjee, G., Sarkar, U., Das, S., Ghosh, I.: Artificial intelligence in agriculture: a literature survey **7**(3) (2018)
2. Beloev, I., Kinaneva, D., Georgiev, G., Hristov, G., Zahariev, P.: Artificial intelligence-driven autonomous robot for precision agriculture **24**(1), 48–54. https://doi.org/10.2478/ata-2021-0008. https://sciendo.com/article/10.2478/ata-2021-0008
3. Eaton, R., Katupitiya, J., Siew, K.W., Howarth, B.: Autonomous farming: modeling and control of agricultural machinery in a unified framework. In: 2008 15th International Conference on Mechatronics and Machine Vision in Practice, pp. 499–504. https://doi.org/10.1109/MMVIP.2008.4749583
4. European Parliament: Directorate General for Parliamentary Research Services.: Precision agriculture and the future of farming in Europe: scientific foresight study. Publications Office. https://data.europa.eu/doi/10.2861/020809
5. Farhud, D.D., Zokaei, S.: Ethical issues of artificial intelligence in medicine and healthcare **50**(11), i–v. https://doi.org/10.18502/ijph.v50i11.7600. https://www.ncbi.nlm.nih.gov/pmc/articles/PMC8826344/
6. Goldfain, B., et al.: AutoRally: an open platform for aggressive autonomous driving **39**(1), 26–55. https://doi.org/10.1109/MCS.2018.2876958. Conference Name: IEEE Control Systems Magazine
7. Jeon, C.W., Kim, H.J., Yun, C., Han, X., Kim, J.H.: Design and validation testing of a complete paddy field-coverage path planner for a fully autonomous tillage tractor **208**, 79–97. https://doi.org/10.1016/j.biosystemseng.2021.05.008. https://www.sciencedirect.com/science/article/pii/S1537511021001082
8. Jiru, J.: Autonomous driving - fraunhofer IKS. https://www.iks.fraunhofer.de/en/topics/autonomous-driving.html
9. Kneip, J., Fleischmann, P., Berns, K.: Crop edge detection based on stereo vision. In: Strand, M., Dillmann, R., Menegatti, E., Ghidoni, S. (eds.) IAS 2018. AISC, vol. 867, pp. 639–651. Springer, Cham (2019). https://doi.org/10.1007/978-3-030-01370-7_50
10. Kriebitz, A., Max, R., Lütge, C.: The German act on autonomous driving: why ethics still matters **35**(2), 29. https://doi.org/10.1007/s13347-022-00526-2
11. LeCun, Y., Bengio, Y., Hinton, G.: Deep learning **521**(7553), 436–444. https://doi.org/10.1038/nature14539. https://www.nature.com/articles/nature14539. Number: 7553 Publisher: Nature Publishing Group
12. Mark, R.: Ethics of using AI and big data in agriculture: the case of a large agriculture multinational **2**(2), 1–27. https://doi.org/10.29297/orbit.v2i2.109. https://www.sciencedirect.com/science/article/pii/S2515856220300110
13. Mohr, S., Kühl, R.: Acceptance of artificial intelligence in German agriculture: an application of the technology acceptance model and the theory of planned behavior **22**(6), 1816–1844. https://doi.org/10.1007/s11119-021-09814-x
14. Muhammad, K., Ullah, A., Lloret, J., Ser, J.D., de Albuquerque, V.H.C.: Deep learning for safe autonomous driving: current challenges and future directions **22**(7), 4316–4336. https://doi.org/10.1109/TITS.2020.3032227. Conference Name: IEEE Transactions on Intelligent Transportation Systems
15. Rohleder, B., Krüsken, B., Reinhardt, H.: Digitalisierung in der landwirtschaft 2020 | studie 2020 | bitkom e. v. https://www.bitkom.org/Bitkom/Publikationen/Digitalisierung-in-der-Landwirtschaft-2020

16. Göggerle, T.: John deere: Autonomer traktor jetzt serienreif - kommt er schon 2022?   https://www.agrarheute.com/technik/traktoren/john-deere-autonomer-traktor-serienreif-kommt-schon-2022-589025

17. Uddin, M., Chowdhury, A., Kabir, M.A.: Legal and ethical aspects of deploying artificial intelligence in climate-smart agriculture. https://doi.org/10.1007/s00146-022-01421-2

18. Rübcke von Veltheim, F., Heise, H.: German farmers' attitudes on adopting autonomous field robots: an empirical survey **11**(3), 216. https://doi.org/10.3390/agriculture11030216. https://www.mdpi.com/2077-0472/11/3/216. Number: 3 Publisher: Multidisciplinary Digital Publishing Institute

19. Vrochidou, E., Oustadakis, D., Kefalas, A., Papakostas, G.A.: Computer vision in self-steering tractors **10**(2), 129. https://doi.org/10.3390/machines10020129. https://www.mdpi.com/2075-1702/10/2/129. Number: 2 Publisher: Multidisciplinary Digital Publishing Institute

20. Weale, A.: Ethical arguments relevant to the use of GM crops **27**(5), 582–587. https://doi.org/10.1016/j.nbt.2010.08.013

21. Wessling, B.: Are farmers ready for autonomous tractors? https://www.therobotreport.com/are-farmers-ready-for-autonomous-tractors/

Author Index

J. Bayer and C. Grimme (Eds.): *Code and Conscience*, LNAI 14400, p. 157, 2025.
https://doi.org/10.1007/978-3-031-52082-2